THIS VIRTUAL UNIVERSE
OF SPIRIT, SCIENCE, AND SENTIENCE

A Personal Perspective on the Nature of Existence

David MacClelland

This Virtual Universe of Spirit, Science, and Sentience
Copyright © 2021 by David MacClelland
Editor – Hannah MacLeod

All rights reserved. No part of this publication may be reproduced, distributed, or transmitted in any form or by any means, including photocopying, recording, or other electronic or mechanical methods, without the prior written permission of the author, except in the case of brief quotations embodied in critical reviews and certain other non-commercial uses permitted by copyright law.

Tellwell Talent
www.tellwell.ca

ISBN
978-0-2288-5201-8 (Hardcover)
978-0-2288-5200-1 (Paperback)
978-0-2288-5202-5 (eBook)

"We shall not cease from exploration,
And the end of all our exploring,
Will be to arrive where we started,
And know the place for the first time."

T.S. Eliot (1888-1965): "Little Gidding" (from the last of his Four Quartets)

★★★

Dedication

To Kenzi, who shared the essence of this journey,

and to my wonderful children, Juliette, and Alexander –

Cherished gifts from the Universe.

Table of Contents

Part 1: Spirit 1
    The Quest 3
    Insights into Meditation 19
    Awakening Mind 45

Part 2: Science 67
    The Human Condition 69
    Influencers of Mind 87
    Inscrutable Reality 111
    Matters Physical 129
    Contemplating Origins 149

Part 3: Sentience 173
    Primordial Sentience 175
    Mind over Matter 197
    Cosmic Influence 217
    Agency of Sentience 237
    Purpose for Being 261
    Resolution 279

Overview

This book is an exploration of spiritual, psychological, and scientific perspectives on the nature of existence and how they are drawn together into one holistic theme.

From spiritual experiences gleaned through meditation and contemplation, Part 1 describes the author's personal quest into concepts of God and sentience, leading to more secular insights into the nature of existence.

Part 2 examines the involvement of psychology and science on factors that can influence or distort our perceptions of reality and insight, and reviews some of the under-appreciated aspects of the physics of our material existence.

Finally, Part 3 imparts an understanding of how fundamental physical and spiritual forces could combine to guide and encourage the universe to evolve from raw unconditioned consciousness into a complex whole, supporting life and sentience; and in which sentient beings play a co-creative role.

Notes to Reader

The "I" word

When did "I" become so important? One aspect of the grammatical style used in this book needs explaining at the outset. While contemplating the subject of ego and humility, and since writing about my own subjective experiences from the first person's viewpoint, I became increasingly sensitive toward the capitalization of the letter "I" when using the first-person pronoun within a sentence such as this one. It is an English grammatical rule to capitalize "I" wherever it appears; however, the continuous capitalization of it in mid-sentence seems to me to be both jarring and conferring unnecessary status. Casting my eye over a few European languages, je, io, yo, ich, for example, I saw no other occurrences of this. When such first-person reference is unavoidable, it felt preferable to use lower case to minimize the visual impact and implied importance of references to the self. Therefore, while an upper case "I" will still begin sentences following conventional grammatical rules, i shall use lower case in mid-sentence when referring to the first person. I do hope this does not detract from this work. Getting used to the lower case "i" might take a little effort at first, though after a while, you might find, as i have, that sentences read more smoothly.

Spirituality

Noticing how frequently words based on the root word "spirit" were used i realized that they might mean entirely different things to different people. To many, "spiritual" may have a narrow theistic definition involving matters of worship, church, religion, and a god. To others, it may conjure phantoms or psychic séances. The use of the word here embraces the hidden or non-physical influences and metaphysical philosophies that appear to have some essential association with our existence, both individually and on a broader scale, and may even spill over into psychological aspects. At an individual level, spirit describes that ineffable, mysterious spark of life that seems to animate sentient existence. "Spiritual" therefore represents a mystically biased perspective that includes metaphysical acknowledgement of hidden influences on existence. Within this broader definition, for example, Buddhism would qualify as a spiritual tradition, although it is secular with no requirement for churches, worship, or gods. Herein, i frequently equate the notion of an individual spirit to the concept of the psyche, which suggests a more secular interpretation.

Analogy and Metaphor

Describing a profound and unfamiliar subject using familiar terms stretches the ability of language. The challenge is in the attempt to ascribe a deeper meaning to a commonly used label, which extends it well beyond the presumed, superficial understanding of it. I am thinking of the fundamental significance of quite familiar yet profound words such as consciousness, spirit, ego, energy, time, and reality. Most of us think we understand the meaning of these words, though when pressed to articulate an accurate description or definition, it turns out we do not. Our language, mathematics, scientific formulae, computer simulation programs – even our thoughts – are all tools for expressing ideas and make use of symbolic forms of analogy or metaphor to model the subject. Each is reliant on common, culturally influenced points of view to enhance a specific aspect of understanding – perhaps to appreciate some less apparent characteristics of current interest or to introduce completely novel concepts. The use of analogy and metaphor here attempts to establish a meaning with which the mind is familiar to create a comprehension of something that it is not.

Introduction

The initial purpose for writing this work was to sort out the confusion of thoughts, insights, and understandings arising from what began as a theistic-oriented spiritual investigation – a personal challenge of seeking the identity of the "real" God. That evolved into a broader, secular appreciation of the nature of existence, the universe, and our role within it. Informed by contemplative meditation, intuitive insights, and a pale understanding of science, psychology, and metaphysics, i pursued a multi-disciplined journey of discovery, and developed a recognition of our interdependent physical and psycho-spiritual nature. I realized that one inconspicuous thread runs through it all – the ubiquitous presence of consciousness. With the practice of meditation and contemplation, i felt i had tapped into a reservoir of partial comprehensions – glimpses into an inventory of metaphysical insights. These instances seemed chaotic and incomplete, although they were becoming numerous. I resolved to document these sporadic insights, to push the boundaries between the describable and indescribable, and to develop a unified description of the workings of existence – in retrospect a somewhat ambitious goal! This personal essay-turned-book attempts to portray that.

While developing my notes, i came to realize that such a record of the combining of contemplative insights with those of science in exploring the esoteric, the nature of existence, and consciousness could also have value in encouraging others to explore the metaphysical for themselves – to find their personal truths and relationships with existence. Indeed, i felt compelled to share these insights, not from a perspective of prescribing how things are, though they may seem that way to me, but to encourage you to find your truths. "Yours," mind you – not second-hand presumptions gifted or imposed by self-appointed spiritual leaders and institutions, or pronouncements interpreted from ancient multi-authored, multiply-edited scriptures. Each of us has the ability, the inherent right, and dare i say it, the responsibility to establish direct communion with our inner self, our psyche, our spirit, and thus encourage our self-awakening. We should not surrender that right to the will and dogma of others.

For one with a strong engineering bent, it was a challenging quest first to experience and then write about such an intangible, partly spiritual, undoubtedly personal, and often quasi-rational subject. A subject that turns out to be, on one hand, smaller than a dot, and on the other, larger than the universe; one that can evoke prejudice, reaction, and even superstitious fear – yet also wonder, awe, and love; and one that remains forever subjective. Thus, was this book born, intended not as a scholarly dissertation, but as a record of my experience – an extended personal essay recounting spiritual discovery, interwoven with contemporary science-based conceptual understandings.

Despite the efforts of prophets and subsequent religions over the ages, our psycho-spiritual or metaphysical evolution has remained relatively stuck – impoverished, even – and by any definition of balanced development, seems lacking. From the outset of this investigation, i determined that the discovery phase of the journey would not involve researching the past beliefs or opinions of others. Therefore, with help from a true friend, i found my way – not outward to an ashram in some exotic Eastern land – but inwards, to introspective contemplation through meditation. Once the practice of meditation became established, much of what insights followed seemed self-generated – in conventional terms anyway. Intuitive metaphysical understandings were uncovered through interior pathways, and once found, i realized they were not new, nor ever secret, but were always already available – to all. From meditative contemplation eventually sprang a deeper awareness, a new perspective on the nature of reality and our purpose. I found that although formal meditation may sometimes provide such awareness directly, it is frequently the attitude and state of mind fostered by the regular practice of meditation that provides an extended receptive state of being. This state enables unsolicited sporadic glimpses of enlightened understandings – and the recognition of them when they happen.

When first started, meditation is a discipline. Once it becomes familiar, it develops into a form of being. Accompanied by authentic intention and humility, it is a crucial tool for subjective enquiry and can yield a whole spectrum of inner experiences. At the superficial end, one finds mental and physical relaxation and rejuvenation. At the more profound end can be experienced a gradual reduction of psychological attachment to the world of material things and conditioned thoughts – a transitioning of awareness, until one is conscious of nothing tangible holding reality together. At the same time, there is a sense of connectedness with everything, of being everywhere – inside and out. This feeling of completeness defies concise description; it is one of being whole, being everything – and nothing.

This book strives to offer secular encouragement to seekers who may be weary of conventional religious fare and are contemplating a similar spiritual enquiry. I have in mind individuals, curious by nature, who feel there is more to life than can be explained by either religion or scientific materialism. Such individuals may perhaps:

- have some knowledge of physics and Eastern philosophy concepts, and want to deepen their thinking,

- be rational but out-of-the-box thinkers, able to mentally explore beyond conventional limits,

- have a keen interest in furthering their knowledge on this subject, or

- have a desire to read about the experiences of another on a similar path.

My concern for conciseness and my limitations of comprehension may have led me to take a few abbreviating liberties with some of the scientific explanations, for which i apologize. Though not intended as a work of reference, i hope the book might offer a sounding board with which to compare your own experiences; and some commiseration with frustrations you may feel as your journey unfolds. Those who have already trod this path may give a nod of recognition to descriptions of familiar scenery as revealed through different eyes.

In the present day, our ability to conceptualize scientific, metaphysical, psychological, and even spiritual insights into more intelligible terms continues to improve, as compared to times when the world's major religions, which still form an influential part of our culture, were born. Application of this current level of conceptual ability to mystical subjects has encouraged a surge of more secular, spiritually open-minded people to give voice to their understandings, discoveries, and philosophies. To that tentative throng, i humbly add my insights and musings, to integrate critical perspectives on existence, the universe, and ourselves.

PART 1

Spirit

*Spirit is not born, nor deceases ever,
Has not come from any, or from it any.
This Unborn, Eternal and Everlasting
Ancient is not slain, be it slain the body.*

Katha Upanishad, circa 900–600 BCE

THE QUEST

The Question

The tension in my stomach told me i would rather be elsewhere. The queasy feeling of stage fright developed as i struggled to find the words to express my response to the question just asked of me. The expectant silence of the room grew, as did my tension. Others had spoken, but they were not my words, did not reflect my thoughts, and offered little guidance to me. Now it was my turn to speak. As the only male student in the class, i already felt emotionally disadvantaged in trying to communicate thoughts from the heart. Probably also the sole engineer in the room, i would likely be a minority "left-brain" thinker among the dozen or so other students – all seemingly more right-brain-directed than i – who were also attending this introductory class at the local meditation centre. The task of translating a heartfelt, though barely recognized thought into a concise response to inform the group, was challenging me – a bit like assembling a sentence in an unfamiliar language. In dealing with these feelings of confusion, estrangement, and inadequacy, i felt uncomfortably outside my element. The innocent question the meditation leader had just asked was, *"why are you here?"*

As often happens, when under the pressure of time, time obligingly dilates. Some of the students in front had half-turned toward me to see whose turn it was to respond, and, i think, as a gesture offering both interest and reassurance. Faces floated into my peripheral vision – every one female, though not all pale. I had noticed earlier that several of the students had dark complexions, perhaps of East Indian heritage. Two women, sitting together in front of me on the spartan, grey-padded, tubular steel chairs, wore striking, crisply pressed, brightly coloured saris and lacy headscarves. They remained facing toward the front of the stark, white-painted, fluorescent-lit classroom, quietly studying the floor. Another wore a simple white sarong, with her black hair exposed in a severe bun. The leader was standing at the front of the room, next to a table draped with a white cloth. On one side of the table, some flowers were set in a huge vase. A framed monochrome picture of the peaceful-looking face of an elderly man, perhaps East Indian, was on the other side. Over the centre, hanging on the wall, was a sizeable unframed graphic, mainly in soft shades of warm pink and red, of rays emanating from a white dot at its centre. Although an unusual setting for me to be in, i sensed it not the least bit threatening. As a newbie, i felt relatively comfortable there.

Calmer now, and while the faces of students and leader remained frozen in this glacial moment, i entertained that question again, more luxuriously. I could feel my mind examining it in different ways, from different angles. Why *was* i here? The enquiry was a simple, general-purpose icebreaker, but, depending on perspective, could be interpreted and responded to at any number of levels. At the superficial level, i was here because it was seven o'clock, and that was when the meditation class in which i had enrolled was due to start. Far from the trivial level, at a deep philosophical level, why were any of us here – here on this planet, in this universe – and did this "why" embrace a god's role, or did it stop at the door of scientific materialism? It was more this end of the question's spectrum of meanings that i was here to explore. Here was a hall of mirrors, though. In looking deeply into one question, you saw reflected in it another, and in that one was reflected the reason for the first. I stifled the mischievous idea of replying that i was here to find the answer to the question asked.

Quest Begun

Although motivated to seek training in meditation and spiritual awareness, i had not come to this meditation class lightly. True to my engineer mentality, and with the help and encouragement of a dear friend who had previously attended them, i had done some research into the organization behind these classes, which had even included interviewing one of the class leaders, though not the one directing this class. The session was one of many free evening classes offered by the local meditation centre of the Brahma Kumaris World Spiritual University, which encouraged self-development through meditation and spiritual focus as a means of improving the community to which one belonged. Brahma Kumaris is a global spiritual organization run mainly by women – an early form of virtual University founded in old India, just before the separation of Pakistan. Its administrative centre is in London, England; the spiritual headquarters are on Mount Abu in India; and many associated meditation centres, such as the one i attended, are scattered throughout the world.

I suppose that was one reason i was there – this serious organization altruistically offered to help anyone's own efforts to develop their spiritual self. There were no strings attached – no fees and no conditions. True, they did it their way; and i was to find some of those ways hard to accept, although i also found that one was always free to question, develop and evolve, based on one's inner voice – once one got to hear it. Brahma Kumaris seemed to offer a somewhat unusual opportunity – one that would help ease my way into a quite different world from the one i had lived in up until then. It provided a spiritually inspired, albeit temporary, haven, where compassion, unconditional love, peace, and just Being were practised. These values were assigned a much higher priority than in the material world where personal gain, prejudice, ego, and being-seen-to-be-doing were more the norm.

The leader's patient face remained frozen in time, with eyebrows arched in a pleasant though questioning demeanour. The clock on the wall behind me uttered a low-pitched, drawn-out tick, sounding more like a cluck. Nothing and nobody had moved in the few milliseconds since she had first asked the question – "why are you here?"

Part of the reason was to find out, if there were a God, what roles He might have in my life. Not that God was unbeknownst to me before this day. We had been introduced, in somewhat abrasive fashion, in England many decades before, when, having attained school age, though born of agnostic parents, i was inexplicably sent to a religious-themed school for boys. In this place, He was forced upon me as a wrathful, vengeful God who i had better darned well obey and worship daily – particularly Sundays – or i would end up in a fiery hell, being tortured by Satan. Furthermore, in case those colourful prognostications for the future of my soul were not sufficient incentive, i would be physically punished in the present world whenever i was perceived to be transgressing from the "one true path" to salvation. In my case, that path was the Protestant religion. No doubt, those raised in other religions could recall similar experiences and assurances – each faith adamantly proclaiming itself as the one true path.

Formative Years

My school, as well as being Christian – involving daily chapel-going, religious studies, and much discipline – also had a strong military training component. At a minimum, one whole afternoon of each week was devoted to training all students to be soldiers, aspiring to be officers. This process seemed to involve an awful lot of marching and military drill – or square-bashing as we called it. On the upside, we did get to use all sorts of exciting military equipment, including various types of communications radios, and working guns of all kinds – from handguns to artillery – and occasionally using live ammunition. A select few of us, who were found to be accurate shots with a rifle, became designated as "Marksman," meaning that we would be better at killing God's enemies at a distance than most. This elite group was provided with copious amounts of live ammunition, considerably more rifle-range time, and often bussed out of school to attend various military shooting competitions, all the better to hone our sniper talents with constant live practice. A Marksman could elect to participate in extra range practice, which excused him from many extra-curricular school activities, such as cricket – which i did not care for – but alas, religious studies were firmly ensconced into the standard curriculum, and non-optional. The incongruity of spending a regular portion of the week devoted to meeting God's sacred needs and commandments, Christian duties in chapel, and religious studies; and a similar period learning to kill our fellow man, was not entirely lost on me. However, whenever i raised this apparent paradox with the wise, i was assured that anyone we killed in wartime would be godless since God was on our side, not theirs, so this mitigated the problem. Even at my tender age, i sensed this argument to be somewhat flawed, yet i was also made aware that to persist in such moral questioning would be actively discouraged. I find it amazing that such a flimsy and partial argument remains in use today by people, religions, and nations on both sides of conflicts.

Under the burden of such powerful, though not particularly charitable persuasion, i did my best to adapt to the compulsory religious dogma and ritual, as set by those self-appointed guardians of my spiritual well-being and belief system. I also discovered it to be true that on those occasions when they perceived me to be transgressing from the formalities necessitated by the one true faith, i would indeed be punished – physically – with the aid of cane or gym shoe. Whether my soul, on those occasions, also felt the wrath of God, i was never able to determine, although i sincerely hoped not. Under such duress throughout my school years, i largely conformed to the minimal required religious norms in a passive-aggressive kind of way. Whenever i did think about it, though, that version of God never sat well with me, and as soon as i finished school, i left God-worship to those who seemed to extract some comfort from the process.

I do not mean that i became a godless person, though; just that i was not enamoured with organized religion as a means of establishing a profound personal spiritual understanding. Perhaps somewhat cynically, it seemed to me that embracing religion was often more to do with evading self-responsibility, building social connections, and attempting to insure against death and disaster, than it was about seeking true spiritual enlightenment. In my mind, there was a significant gap between religion and spirituality as a means to resolve my innocent questionings and embryonic intuitive understandings. I have since come to believe this gap may be an even more substantial obstacle to self-development than the more obvious differences between spirituality and science, in resolving personal spiritual questions.

The idea of a god did intrigue me, though – the "real" God, that is; not the fearsome one i left at chapel. I felt that there had to be some single-point explanation – such as a god – for the existence of rich and varied sentient life, non-sentient forms, and natural events that occur in our world and the universe at large. For me, existence and sentience had to result from more than just a mindless sequence of random evolutionary events occurring over a long though finite time.

Questioning Phase

From time to time through my adult years, i would think childlike thoughts about God. Why would He want to be worshipped, anyway? Granted, the way it had been taught to me, He had performed a fantastic feat in creating the universe and us – all in six days – but i could not believe that he would then crave endless applause, like some insecure magician. If i had just created an army of ant-like creatures in my image, would i need them to line up daily, and for two hours on Sundays, to worship me? What kind of obsessive ego-pathology would that imply? Intuitively i knew that the "real" God, if He were person-like, was not someone who demanded recognition, nor discriminated against anyone for not being on the deemed "right" path of devotion. Nor was He a being to be fearful of – perhaps more like a gifted friend to hold in some awe.

As for His management style, i could not envisage Him as someone who had to have hands-on control of events and people – responding to individual prayers and deeds, like some Santa Claus checking his list twice, and meting out arbitrary favours and punishments. I could, however, envisage Him as the One who had kick-started the whole shebang, and was now lovingly watching His creation – perhaps from a distance – as it unfolded on its own, in the form of a slightly wobbly, self-organizing system.

Then again, maybe paradoxically, i was always uncomfortable with the implication that God had to be distant, to stand outside our system of existence in order to be omnipotent, omniscient, and all the rest. Why would He have to be remote – intangible, untouchable – to be perfect? Such distant absolute perfection seemed to confuse my admittedly fuzzy perception of Him. Perfection, it turns out, is relative.

In my musings, it seemed that our self-imposed distancing from God's perfection has led to the gradual deification of Jesus Christ, who started out being just an ordinary person. From being a remarkable individual, and most likely an enlightened one, gradually he has been transmogrified to some degree by the church, from being a moral mortal man, to the son of God, to the *only* son of God, to a surrogate for God, and thence to God Himself. In the Catholic Church, there seems to be a trend to do the same thing with his mother, Mary. Not much mention of Joseph though; the idea of two fathers being perhaps too embarrassing for those conservative authorities who, centuries later, felt it necessary to invent the virgin-birth concept as a revisionist solution to that moral issue.

Even in this age, these kinds of questioning thoughts may seem heretical to some. However, to me, they were very innocent, authentic questionings of an institution's flagrantly flawed assertions and superstitions. For me, organized religion was a very confusing, prescriptive, prejudicial, and irrational subject. Overall, i considered that it probably had some positive social value, and undoubtedly some negative social contribution, though for advancing one's spiritual awareness, it seemed neutral at best.

Then, there was the image thing. Despite the mute assurances of countless religious artists, mostly of European extraction, i felt uncomfortable with God's image being a permanently autocratic, or fatherly – depending on the artist's mood – older man in a white robe with a white beard and Caucasian features. Was He ever young? Would He get even older? If so – heaven forbid – would He die? Was Jesus effectively His understudy? His replacement? And why a He? Endowed with such compassion and love, albeit conditional on our following the Church's strictures, why not a She? The possibility that God should be female would certainly be contrary to monotheistic religious history – of

*Man*kind, mind you! Why should God be *any* gender, why not an "It"? However, "It" sounded even more remote, so maybe It should stay as a Him – for now. The portrayals of God to which i had been exposed spoke more to the desires and deficiencies of the artists, authors, and promoters than to any notion of reality. This God, as taught to me, was a conditional, metaphoric God. If the publicly perceived God was but a metaphor, perhaps i needed to find the real Him myself, and not rely on the self-serving accounts of others. I concluded that one owes it to oneself, and any acquired god, to find spiritual awakening firsthand.

Considering metaphors of spirituality reminds me of an experience i had in my childhood when i was eight or ten years old. My parents had taken me to an exhibit - a rigid-walled tent in the midway - at a seaside fair. It was billed as a "Camera Obscura," which i learned is Latin for what we would call a pinhole camera. The difference with this camera was its size – you could walk inside.

Inside the darkened room, once our eyes had become adjusted to the dark, the operator would uncover a tiny aperture in one wall. There were no lenses, no electronics, just a pinhole. On the opposite wall was a white screen. A coloured image of what was happening outside the room projected onto it through the tiny pinhole. When the sun was shining outside, the picture on the screen in the darkened room was quite bright. When clouds came over the sun, the image faded to a more ghostly intensity. A woman walked her pram across the screen, and, just like the kid i was, i ducked out the tent door to check on reality. Yes, with my eyes tearing up from the bright sunlight, i could see a woman was there, walking a pram adjacent the side of the tent with the now invisible pinhole in it. I was impressed.

I darted back inside. After my eyes recovered from the brilliance of the reality outside, i could once again discern the pale images of it on the screen. If i turned my head to determine the source of the projection, all i could see was a tiny point of exceedingly bright light at the pinhole in the wall opposite the screen. Nevertheless, i knew that if i could somehow squeeze myself through that pinhole, i would emerge into a reality more brilliant than anyone could imagine from within the tent. Perhaps, i thought, God and our souls live in the vibrant, dazzling metaphysical reality, while we – our physical bodies and the material universe as a whole – are but pale projected images. For me, it has become something of a personal metaphor for the spiritual side of life. That experience had triggered in me the same sort of thoughts about reality, upon which, as i would learn very much later, Plato had mused. His metaphor was a campfire at night at the mouth of a cave, and the cave dwellers trying to make sense of the outside "real" world from the flickering shadows that the fire projected onto the cave wall.

Discouragement

During my high school years, supported by an inherent mechanical aptitude, an intense curiosity about how things worked, and encouragement from my parents, i had developed a love for physics and mathematics, and a subject called applied mechanics, which was primarily theoretical engineering. I would spend much of my free time playing with concepts i had read about or learned in class. Once, after a physics lesson on the nature of light photons, and armed with a rudimentary understanding of gravity, i came up with the concept of a massive star, much bigger than our sun. It would be a place of such high gravitational forces that even light could not escape, and i concluded it would therefore be invisible. Furthermore, something like it might form the undetectable gravitational centre of our galaxy and possibly the universe, holding them all together. That evening, i enthusiastically wrote some notes and diagrams about it – which i still have – and with eager anticipation, took them to the school physics master for discussion the next day.

He looked at my notes briefly, and then crushingly told me the idea was nonsense. Unfortunately, at the time, i accepted his judgment and did not pursue it. Little did i know that this same concept, though still controversial, was already being considered among astrophysicists of the time, though not by my physics teacher. Not until a decade or so later was the phrase "black hole" introduced to the public as a popular science term for describing such a theoretical event. Subsequently, along with much more recent credible physical evidence, its role in holding galaxies together has also become a commonly accepted theory in mainstream cosmological science.

Despite such a discouraging setback to my enthusiasm, i did well in physics and mathematics in my school final exams, and later in electronic engineering at university. I still consider that, had i not gone straight into the engineering industry when i graduated, i might well have found my way into theoretical physics. However, in my chosen professional life, there was little overlap between ideas about the workings of the universe and producing things that work in this world. Nevertheless, i was always interested in learning more from the fields of physics and astrophysics, particularly at the micro- and macro-scale extremes. Through the popular-science press, i would keep my understanding current on concepts, ranging from the big bang, black holes, and relativity theory to subatomic particle physics – in particular, the intriguing, novel, counter-intuitive, and somewhat coy revelations of quantum-mechanical theory.

The Questions Return

Much later in adult life, the intrigue generated by concepts of God returned and grew stronger. Having navigated a half-century of being, and in the aftermath of some emotional trauma, i had decided to take stock of my philosophy of life and look hard for the "real" God. I felt that perhaps ignoring the spiritual aspects of my life might be causing my emotional "problems." Indeed, i even found myself profoundly questioning what or who i was. These may be classic symptoms of the popularized "mid-life" crisis, but they can also be indicators portending the onset of personal change. Since graduation, concepts of God and the Universe had occasionally visited me in my thoughts and dreams. I had maintained my interest in the field of physics throughout my life. I even tried incorporating hybrid perspectives of both scientific and spiritual ideas into various attempted novels, which, without exception, i aborted after the first few chapters.

Nevertheless, i became more conscious of a possible convergence between scientific and spiritual understandings – a bridging of physics and metaphysics. With an open mind – it seemed to me – one perspective might shed light on the other. The specific moment that crystallized this insight occurred overnight. I had been dreaming of these hybridized perspectives on concepts describing existence and awoke with an authoritative poem fully established in my conscious mind. While not having the clarity of a "Eureka!" moment, this spontaneous dreamed poem would mark a renewed active interest in developing my spiritual side further.

Fortunately, it was a Sunday; in retrospect, the delicious irony seems appropriate. Without needing to go to work, i leapt out of bed to sit at my computer for most of the morning, transcribing a record of this dreamed poem before it could dematerialize from conscious memory. Although i lost a small part of the end of the dream and had to patch part of the last stanza consciously, i was astounded at what had come to me. The expression did not seem part of me – at least not the me i thought i knew. From the perspective of an engineer, what i had written was quite incongruous – way, way out. The poem, while not being particularly fine poetry, spoke with a voice of authority on matters profound. Much later, i titled it "Genesis," and it concludes this chapter. Maybe it would have been run-of-the-mill for some, but it meant a lot to me. Its spontaneous creation left me bursting with feelings of awe, amazement, enthusiasm, accomplishment, and serenity, among other emotions. The experience may or may not have been a peak one; it was undoubtedly pivotal.

Subsequently, it dawned on me that embedded in my future life's path was the desire to understand more fully that poem, that spontaneous insight. However, to do that, i needed to embark on a journey of spiritual understanding. I recalled an adage about walking in a featureless desert – that both left and right paces should be equal, to avoid returning to the same spot. Until now, my life, like those of many others, had predominantly favoured the materialistic pace. Now, my materialistic and spiritual strides needed to become more in balance, or else my life's journey might continually circle back to the same unsatisfactory place.

Intuitively, i felt that the spiritual-quest formulae of studying the world's religions, poring over scholarly books and scriptures, undertaking pilgrimage quests to faraway places, or becoming a monk, were not going to be my way of pursuing the profound personal answers i sought to my as-yet-undefined questions. I had already surprised myself once – that poem from within – and i sensed further revelations also lay within me, as i now know they lie within each of us. Meditation, i thought, might be the key to accessing that knowledge – but how to start.

So, why are you here?

Even dilated time must pass eventually; the leader's expectant face reminded me that i needed to say something in response; something intelligent, something pertinent, something authentic, and something soon. Why was my brain having such difficulty framing an answer?

I suppose because i was taking the question so seriously. Still, i felt that under the anxious surface, the answer was already waiting, fully formed; all i had to do was access it. However, instead of concentrating on that task, my mind had been distracted, pondering the make-up of the class, the decor of the classroom, and my spiritual history. Was this avoidance, denial, or what? I reminded myself that this was not a business environment; i could, to use a colloquialism, let it all hang out – none of the reasons voiced by other attendees resonated with my emerging feelings on this subject. I needed to draw out and define my thoughts, then verbalize them. Why *was* i here attempting to learn the art of meditation? What was it that i thought it might help me do or discover? What was driving me to participate in a disparate group of people so different in their thought processes from my day-to-day friends and peers? Had i thought this through? A realization blossomed within me that: yes, unconsciously, all my life, i had. The feeling in my stomach quite suddenly relaxed into a warm glow of confidence. I calmed my thoughts to voice them from the heart, and not to sound like a carefully orchestrated business response. Just as it comes, blurt it out and let the class deal with it!

Time resumed. I responded in a gush, *"I want to understand who God is and how He made the universe."* As i finished, i gave the class and the leader a modest, self-effacing grin to show that i realized this was perhaps rather a tall order – some two dozen appraising eyes, some empathizing sighs, then solitary expectant silence. The leader, Sister Karen, after the briefest of hesitations, acknowledged the depth of this goal with a serene smile and moved on to the next student. At this small meditation centre, i had started – the first baby step – on my quest for spiritual truth through the learning and practice of meditation.

Genesis

I am as old as the universe. I was born of white heat.
At that point in time, my soul existed; my body could not.
The magnificent outrush of pre-atomic elements was part of me.
The light was my blood, the heat my flesh.

I witnessed the coalescence of those elements that would form all atoms.
I was part of that excitement, part of that imperfect apocalypse.
Many others were there; I was not alone.
There was everyone that would ever be.

My body took billions of years to form.
I borrowed matter from beyond the galaxy to make it.
I let my components come to me from the universe; each part had many uses.
Now they are me, a gift from creation.

My soul has always known existence, and always will.
Not physical, for a soul is not made from matter, it has no dimension.
Not fleeting, for a soul has no time.
I am my soul; my body is borrowed from the universe and will be returned.

I am not contained within my body or my mind.
I am associated with it for a fraction of time in a limited dimension.
Impermanence experiencing spacetime.
A butterfly savouring the blooms of life.

INSIGHTS INTO MEDITATION

"Prayer is speaking to God; meditation is listening to Him." Many have quoted variations on this aphorism – including Einstein. Most meditations do not target a god per se, although the pithy point of this quotation nevertheless holds. Prayer often seems to involve asking for help and favours, thus implying a more outwardly directed monologue. Meditation, however, encourages an open, inward-flowing form of awareness, with the subsequent potential for dialogue, enquiry, and learning, requiring neither deity nor prophet.

Perhaps since before the dawn of language and art, some 40,000 years ago, humankind has been intrigued by the insubstantial underpinnings of our perceived physical existence, and the metaphysical, psycho-spiritual explanations of mysterious manifestations at all levels. Variously ascribed to miracles, deities, spirits, magic, chance, and fate, most such mysteries are likely thus enshrouded due to our inability to perceive the mechanics behind the event witnessed. As our society evolves and broadens its technical understanding of the physical, the mystically credited portions of inexplicable events correspondingly shrink. At one extreme, scientific materialism, or physicalism, takes this trend to imply there are no metaphysics; that everything from quarks to consciousness has or will have a physical explanation. At the other extreme, much of humanity continues to rely on outdated, superstitious themes to help explain the incomprehensible. Both views are flawed.

At the limits of our perception, reality seems to intersect with some form of a non-physical enabling state – sometimes referred to as a ground of being. This ground of being alludes to a state that remains transparent to our physical world. We find ourselves in the dilemma of trying to employ rationality and reason to investigate these currently metaphysical aspects of our existence. The materialist view assumes we know what physical phenomena such as energy and matter are, and that we can explain all things in terms of them. However, although we may have learned much about these phenomena, we do not have a robust understanding of them, and so such explanations seem like trying to measure length with a piece of elastic.

Recent hints undermining materialism's largely unjustified assumptions surfaced in the last century with the theory of relativity – which showed gravity to be a modulation and modulator of spacetime – and more recently, quantum theory – with its repercussions on the fictions of solidity, deterministic certainty, and local realism (the principle in physics that assumes a system's state already exists before measurement and can only be influenced by local events). Science is the best method of investigation we have; however, the fact remains that despite creditable advances, many of the latest ideas on physicality still posit the need for metaphysical or non-physical contributions. Examples of this include particles that are formless until observed; additional undetectable spatial dimensions; uncountable numbers of other universes; massive amounts of unknown and invisible mass-energies; super-luminary communication; and the many inexplicable attributes of consciousness. I am not scoffing at these concepts – they represent exciting and challenging areas of investigation – however, even though respectable, and peer-reviewed, they are speculative placeholder explanations for otherwise inexplicable phenomena. Thus, it seems that scientific materialism might have a long way to go to achieve a credible, physical foundation for explaining all that there is – at least one that is not grounded in a faith as tenuous as any religion's.

The alternative extreme – reliance on superstition and theism embroidered with dogma and ritual to explain the mysterious – is generally pre-rational thinking. It cannot endure much rational discussion and exploration before bumping up against self-serving investments of blind prejudice and ego-induced passion, frequently in the guise of faith. Not all spiritual views necessarily fall into this pre-rational category, however. For example, the middle-way school of Buddhism has largely matured out of the superstition trap, and through rational, introspective self-enquiry and reflection has developed a form of psycho-spiritual investigative philosophy and psychology. Potentially, this might well contribute to a more balanced worldview – a subjective-based process for helping understand metaphysical aspects of our reality.

As a species, over the last few millennia, we have made great strides in understanding and even moderating our physical environment in primarily creative ways. Survival and improving our level of wellbeing is a natural, instinctive thing to do, and enhancing pleasure and avoiding pain is a fundamental expression of most biological forms. However, driven by our inherent curiosity and aptitude for discovery, our material advances continue to develop slightly beyond our moral abilities to prudently manage them and their associated unforeseen consequences. Perhaps that is the only way – nature's daring (some might say brutal) way – of advancement and evolution. Push the safe-comfort envelope and, if you survive, you will have evolved. Still, we can see that these material advances are mostly in relatively limited sections of the community of humankind, and even in those, there is some evidence of dissatisfaction with our excessively materialistic-biased advancing state.

Such dissatisfaction can be the result of a residue of unmet expectations – we are doing it to ourselves by projecting unrealistic expectations into our future.

Within our present society, our material focus is not in balance with our non-material awareness. Our outward attention largely denies any inward attention. Maybe that is all it can be for now since our species is still evolving and may yet bring its physical and metaphysical self-development into balance. In the general evolutionary timescale of most other dominant complex species of comparable size – we are not talking fruit flies here – our evolution to date puts us at the young end of a typical species survival timescale. Environmental and self-imposed extinction excepted – a not inconsequential risk – we can anticipate a much longer evolutionary life into the future than we have already lived. In evolutionary terms, we are advancing quickly, perhaps too rapidly for our own good.

To paraphrase an adage attributed to various sources, including Benjamin Franklin and Albert Einstein, although probably anonymous – insanity is defined as continually repeating the same action in the expectation of a different outcome. However, these symptoms can also suggest immaturity, an early part of a developmental learning process. For humanity to mature in metaphysical terms, to develop wisdom or merely retain sanity and survive, the time has come to try something different to resolve inevitable conflicts. Another adage – attributable directly to Einstein – is that problems cannot be resolved by the same level of consciousness that created them. Our thought processes, our awareness must evolve for us to become our own resolution. If a more balanced personal philosophy – including space for meditation and contemplation, authentic presence, a quieted ego, and compassion – were to become the philosophy for life for every member of our species, then peace would surely follow.

Spiritual Traditions

Over the millennia, humanity has sought the psycho-spiritual betterment of self and society, generally through one of two ways. The dominant path in the Mid-Eastern and Western areas of the world is essentially through theistic, belief-in-narrative religions. Typically externalized through ritual and devotionally oriented, it can be either monotheistic or pantheistic. These religions offer the believer, for the price of regular worship, the hope for easement of affliction and a passport to an exalted end-state — provided the believer adheres to their narrative and strictures. Some of them may consider non-believers as excluded from such benefits, punished in this world, and eternally damned in the next. While such prescriptive religions can have beneficial effects on influencing their practicing member's social relationships and morality, they also provide a contrary potential for discrimination, divisiveness, and violence in their god's name.

The traditions dominant in the Far East, and becoming evident in the West, offer psycho-spiritual disciplines of self-enquiry. This approach tends to be meditation and contemplation oriented, and mainly, though not exclusively, secular. The self-enquiry path, for the price of self-discipline, quiet time, and authentic thought-management, offers the potential for awareness, compassion, mitigation of suffering, wisdom, and savouring a measure of enlightenment. Furthermore, Mahayana Buddhism – the Middle Way School – has developed a subjective science around the exploration of the psyche, chiefly centred on meditation and contemplation as the chief investigative tools. Directionally, there seems some overlap between such disciplined spiritual self-investigation and similar aspects of the cognitive, non-physical sciences. Indeed, if one is uncomfortable with meditation as a spiritual tool, then it may be thought of as a psychological one.

Both Western and Eastern paths have in common the intended personal outcome of accruing benefit and avoiding suffering – not an unreasonable objective – although they both have also included their fair share of ritual and dogma in the path to that goal. Perhaps the key distinguishing features between them are that the belief-in-narrative religions espouse faith, worship, obedience, and prayer. In contrast, the self-enquiry traditions choose meditation, self-discipline, mindfulness, and contemplation. Some traditional forms of Buddhism still support dogma, ritual, and even imagery; however, in his book "The Middle Way: Faith Grounded in Reason", the Dalai Lama positions the Mahayana School as making room for rite-free, pragmatic secular understandings, even for the non-Buddhist.

Scholars of pantheism propose that the plurality of gods – which term includes goddesses – represent different facets or characteristics projected by humans. Many devotees feel free to choose a favourite god or goddess of their own as the focus of their personal homage. Monotheistic gods may also display the many characteristics projected on them by the humans worshipping or representing them. Indeed, historical evidence shows that even for the same god in the same religion, significant characteristics of that god alter over the years, as the worshipping society changes.

The goal of the Brahma Kumaris was to enable people to see themselves as pure beings, reinforce their self-esteem, and become self and spiritually aware through teachings, workshops, and meditation. The mental processes and support needed to realize one's profound being are also available from other authentic organizations, such as Buddhism, in which self-realizing meditation and awareness – not prescriptive dogma and ritual – is the central focus. Concepts, perspectives, and philosophies shared between Brahma Kumaris and Buddhism mostly centre on authenticity, awareness of ego, detachment, and moderation of the mental state. The meditation technique taught, which is also effective in Buddhism, is the open-eyed meditation of Raja Yoga.

What is Meditation?

Meditation is possibly the only intentionally directed process that can prepare one for accessing psycho-spiritual awareness and enlightening insights. It can open the mind to supporting a state of awareness and enabling a measure of realization of one's inner self. Meditation and inward contemplation are not confined to matters that may be considered spiritual; they have a much broader application such as helping one become better acquainted with oneself, one's outer world, relationships, reality – everything. They can be the basis of careful subjective enquiries leading to insights while following similar protocols as established by objective scientific principles. They may even provide glimpses into the very foundation of our existence.

Most spiritual traditions consider meditation a laudable practice. For our purposes, meditation involves profound contemplation of spiritual, philosophical, or other abstruse topics, elevating conscious awareness, or self-directed cessation of active thoughts. A key feature necessary in virtually every form of meditation is quieting the mind from both discursive thought and mindless ego-chatter. Such calming requires both the rational mind and the chattering ego to let go. This single feature turns out to be the most challenging aspect of meditation, particularly for our modern-day action-oriented, materialistic monkey-minds. Meditation, in its most basic form, is the undirected or receptive meditation that i describe as open or neutral – just going with the flow. The Zen term for this receptive or open awareness form is *Samatha* – a calm abiding or quieting of the mind. The other mode of meditation, which i call the theme-guided mode, relates to the Zen concentrative or focused attention form, termed *Vipassana* – a clear seeing or systematic mindful investigation into self, intuition, and phenomena.

Theme-guided meditation is more directed, in which the practitioner selects a theme or topic, usually before entering the meditative state. That theme or topic then becomes the sole or main topic of concentrated one-pointed focus for intense meditative investigation. It could be a physical object such as a flower, whose physical presence prompts us to return to that focus whenever our mind wanders. Alternatively, a more abstract topic may be chosen, such as a virtue, a philosophical concept, or even a koan. This form of meditation can be productive and surprisingly energizing. It can develop into a discipline to engage and tire the conscious mind, or to become a form of active enquiry. I initially called such an active enquiry form the "workshop mode" to distinguish it from the neutral or receptive type. During this more dynamic form of theme-guided meditation, what takes place has more the flavour of a dialogue. Sequences of thoughts, clarifying or constructing concepts or information, bubble up on their own, guided by or responding to the general sense of the original topic of enquiry chosen for consideration. The process is hard to describe as anything other than dialogue; one gets the feeling of a question-and-answer session taking place. There are no other parties, voices, words, or visions, only spontaneous responding thoughts, raw understandings, and dawning realizations – vignettes of cognition.

A frustrating characteristic of the meditative state is that one can only allow oneself to be aware of being in it once it is over. If one actively acknowledges being in this state while still within it, the state dissolves, leaving but a wisp of memory. Only when it is over can you attempt to recall the experience from within memory, then consciously bring it into contemplation. Words or images are often inadequate to describe a meditative insight, hence the abundant use of analogy and metaphor. On emerging from meditation, the action of trying to transcribe the recollected insights from one's right brain to the left for further contemplation and interpretation is inefficient. The images and realizations fade and blur in the process, and often what remains is a pale emotive after-image of a just-out-of-reach comprehension. Here is the paradox – the harder you try to brighten that image to clarify the understanding, the faster it recedes from your grasp.

The "workshop mode," where a relatively unstructured question-and-answer realization session can take place, seems to enable some degree of constructive, iterative exchange, the better to grasp a particular understanding. However, the process of interpreting an insight and projecting it from the mind is laden with pitfalls, of which the chief one is the distorting influence that ego can have on the conceptual appreciations of the experience. The fruits of meditation are an entirely subjective experience. No impartial third party can police the authenticity with which the meditative experience evolves into thought, or how that thought becomes communicated. Thus, there is always the potential for inauthentic hyperbole, invention, or denial. Authentic meditative interpretations rely on the experiencer's purity of intention and communicative abilities.

An entirely different experience is a clear-sighted realization, which is something one cannot plan to perform. It just happens, or not – mostly not. It is a possibility that can interrupt any type of meditation or even a contemplative mood. Alternatively described as a lucid dream, luminous mind, or clear-sky vision, it is usually a sudden mindful realization that is so intense that there is absolutely no doubt in the mind of the realizer that a profound truth has been witnessed. The intensity of this sudden interposed experience can make it difficult to avoid reacting – thereby collapsing the meditation – though one comes away from such an encounter with a strong sense of elation.

Clear-sighted awareness can be triggered in meditation, relaxation, or by some purely arbitrary physical or emotional trauma, shock, loss, or even a relatively minor spontaneous occurrence in one's normal reality. Ideally, should such random awakening occur, one would recognize it for what it is – communication with one's spirit or psyche. The resulting recognition might then encourage the recipient to perform further contemplative enquiries to intentionally seek out more understandings. However, in the case of an unprepared mind, a more likely outcome might be not recognizing such random enlightenment at all but to dismiss it as an anomaly, to ignore its gentle tug, or subvert it into some ego-serving distorted version. Far better, then, to have prepared the soil, as it were, for receiving any seeds of awareness, insight, and revelation – whether they arrive by accident or design – and to encourage their manifestation by practising meditation. Still, some may ask, "why?"

Why Meditate?

Having a purposeful reason to meditate is effectively counter-productive. Purposefulness creates expectations, objectives, goals, and judgments. These may then develop into an antithetic atmosphere for meditation, effectively stifling any real progress. Meditators generally surrender to the process; they do not seek goals or strive for outcomes from a session. The enigmatic Star Wars character, Yoda, cryptically counselled young Luke Skywalker, who was struggling to recover his spacecraft from a swamp telekinetically, with *"There is no try, just do!"* As applied to meditation, a paraphrased version of this wisdom might be, "There is no try, just be!" Meditators wish to meditate, to rest in the awareness of being or insightful contemplation, with the emphasis on "being" in a state, not "doing" an activity.

This hyperactive world understands and embraces "doing," but has difficulty comprehending and even condoning taking time out for just "being." That is not a criticism of "doing." Our culture, science and technology would not have advanced to its present state, for better or worse, if all participants were fully engaged in masterful inactivity! Actions are essential to our survival – they get things done – but this commentary represents a call for balance. We can move away from the dualistic mindset where we feel we must choose between the physical and the metaphysical, between the doing and the being. A downside of unrelenting competitory activity is frequently characterized by the absence of peace – for both individuals and society. Therefore, while avoiding purposefulness, let us go through some reasons for meditating.

A well-balanced life requires a measure of both doing and being. Balance is not having to select one over the other. Life's complete set of attributes – its total character – requires both. Our species cannot evolve wholesomely if each day must be justified by being full of doing; we must also learn to maintain, enrich, and explore our silences, to cultivate the gaps between the doing. That is why we still cling, quite rightly, to the idea – though perhaps less the execution – of exercise, weekends, and vacation. However, how many of us forego those small temporal oases of potential de-stressing, for the sake of acquiescing to peer pressures to meet other, more "productive" commitments? Through intentional meditation, we can improve our faculty for awareness – an awakening to the true nature of the environment, our existence, and ourselves. By increasing our awareness of the metaphysical, as well as the physical, we will improve the quality of our complete existential experience – of incomparable value to us and perhaps to the entire universe.

Our materialistic society seems fascinated with the new, shiny, and colourful. We switch from one offering to another like the proverbial kid in a candy store, looking for superficial instant gratification to satisfy subliminally planted unrealistic expectations. With our physical diet heavily influenced by metabolic stimuli of dubious nutritional value, and our mental diet continuously influenced by hyperbolic audiovisual creations and other excitants and combined with the lack of voluntarily induced quiet time, we are becoming a race of sensory-stimulation addicts. We are learning to be inattentive to the profound, to the true needs of others, and, more importantly, to ourselves. Forces within our society encourage individual inattentiveness and social attention-deficit disorders. A commitment to meditation will shout, "Enough!"

In the context of meditation, awareness – the increased conscious visibility of information, thoughts, or feelings manifested in the unconscious – can be expanded and heightened to a deeper appreciation of both physical and metaphysical possibilities such as instinct, intuition, emotion, and the entire psycho-spiritual spectrum of comprehensions. Such appreciation can reduce the degree of grasping for the familiar, and the prejudice, cynicism, and fear with which we might otherwise greet change. It develops a disposition of being open to new inner experiences and insights, enabling a better understanding of oneself and the nature of our impermanent reality.

Consider the frustration and suffering resulting from attachment to unmet expectations, grasping for the unattainable, and investing in an illusion created by ignorance. There is plenty of evidence in our midst that demonstrates how externally-provided portrayals of the expectation of happiness – usually in hyperbolic superficial forms – are frequently unrealistic deceptions. Many try to live the culturally influenced expectations of others – from the choice of cereal to the dreams synthesized by the promoters – most fail. It seems the more prosperous the nation, the more preposterous is this condition.

We know that living life judging ourselves by the yardstick of others is a sure recipe for pain and suffering, and that attachment to unachievable outcomes can be brutal. We know that, yet we all do it to varying degrees – repeatedly. We seem to convince ourselves that such suffering is a temporary phenomenon, but for the last three and a half thousand years, Buddhists have been talking about the Four Noble Truths, which point out that pain, suffering, and unhappiness are caused chiefly by grasping at illusion, and ignoring the true impermanent nature of reality. Those ancient Noble Truths refer to healing those underlying conditions by awakening to truth, the wholesomeness of self-validation, and our true nature; by looking past the synthetic illusion that we had been programmed to believe and seeing authentic reality for what it is. Through meditation, self-enquiry, and contemplation, we can awaken to an authentic reality, thereby choosing a path toward the cessation of suffering. Perhaps the most profound reason to meditate, therefore, is to awaken from ignorance, cease all suffering, and improve one's psycho-spiritual understandings – a sublime purpose, and beneficial when achieved at any level.

Meditation Process

Accessing the condition of mindfulness and enlightenment is not usually a one-time event, but an intentionally progressive development. However, spiritual awareness might also be induced by sudden trauma. Some Buddhist teachers are even reputed to strike their students with their staff when they judge the time as right, to shock the student into spiritual realization. Other unsolicited conditions might also result in similar forms of psycho-spiritual access, such as major stress, physical or psychological trauma, near-death experiences, or just a rare bolt-from-the-blue experience. However, such sporadic and involuntary events are not reliable methods; they are difficult to plan for, harder to repeat, and perhaps most importantly, the associated causal circumstances may be less than desirable.

Chemically induced euphoric states are credited with producing episodes of spiritual awareness, among a range of other "mind-blowing" experiences. However, a significant drawback is that the same chemical that initiated the effect also tends to warp the mental perception needed to authentically interpret and understand the experience, resulting in something less than a clear insight or memory of any actual revelation. Electronic auditory entraining of brainwave patterns causing symptoms similar to those observed during deep meditation are promoted, and may accelerate relaxing and entering the process, but with inconclusive direct results.

Most quick-fix no-effort alternatives to the more traditional approaches to meditation overlook an essential factor. Whatever causes a profound experience, without the mental preparation that comes with practising the discipline of meditation, the affected mind may not be in a sufficiently unconditioned receptive state to cope impartially with the processing of any revelations that occur. An unprepared mind may instead deny or distort the experience.

Howsoever a revelation occurs, a practitioner of meditation is more likely to have a receptive, egoless mind sensitive to the possibility of such unexpected events and be open to the unconditioned contemplation and interpretation of them – even if their occurrence did result from a random event.

Without the discipline of meditation to develop one's sensitivity to an opening into psycho-spiritual knowledge, one's journey into spiritual understanding might become wholly dependent on witless waiting for a random trauma, and were that to occur, expecting it would make sense spontaneously. Unfortunately, this latter approach seems to be the unthinking path followed by much of humankind. Those doing so give up their spiritual responsibilities to this kind of capricious lottery, or, arguably worse, to someone else's preconceptions. As has been said of opportunity, spiritual revelation favours the prepared mind.

The desired goal of meditation is awareness, blessed by some attainment of nirvana – a state of total transcendental bliss and peace characterized by the elimination of desire, suffering, ego-consciousness, and their associated stresses. Achieving such enlightenment means transcending from a base spiritual condition to a more evolved one of realizing universal awareness. In Zen terms – a school of Mahayana Buddhism – it is an ongoing state of unconditioned knowing of our fundamental oneness with the universe (*satori*). An alternative term for enlightened is "awakened," which may be more appropriate for modern-day usage, implying as it does that these profound understandings have been under our sleeping noses all the time. Our awareness just required awakening to them; to see through the illusion of rampant materialism; to pierce the veil of Maya.

How to Meditate

For the beginner, the choice of location can be quite significant, although with experience one will be able to meditate spontaneously almost anywhere. Choose a relatively quiet place, inside or outside, and plan for a minimum of distractions. A like-minded group at a meditation centre with an experienced facilitator can be beneficial, particularly if supported by a learning structure. Subsequently, one can apply these learnings to meditate whenever and wherever one chooses – in a formal group setting, on one's own, or opportunistically in public places.

Minimize external factors triggering the body's senses and thus the conscious awareness of them, such as drafts, drips, flashes, noises, smells, and motion. Particularly if meditating within a group setting, consider the more delicate issues of personal hygiene and bodily emanations. Strong fragrances, flatulence, body odour, halitosis – including that caused by garlic – are generally unwelcome during meditation when they can be significant distractions for one's nearest neighbours and not really in the spirit of compassion. For your comfort, check your bodily needs before starting; these range from appropriate clothing to voiding a full bladder. You may only plan to meditate for twenty minutes, but if it turns out to be fruitful, it is frustrating to have a beautiful experience prematurely collapse, due to your conscious mind screaming that you have to go!

For a meditation posture, the body should be comfortable in both temperature and position. Adopting the classic full lotus position on a hard bench – like the position taken by the monks of Tibet – is not necessary unless you are comfortable doing so. Find a relaxed, dignified posture that helps to relax the mind while also inhibiting sleep. One can choose from many postures – sitting, lying, standing, walking, stretching (hatha yoga poses), or working (karma yoga), among others. The choice is your preference, so experiment with the effectiveness of different alternatives. Note that if meditating while walking or working, it is usually safer to choose the open-eyed style of meditation! The key is still the same, though – be relaxed and stress-free. If you choose a dynamic form of meditation, use fluid, comfortable movements with an economy of exertion.

Quiet, soothing sounds can be helpful. These may be a spoken commentary, the dulcet sound of bells, tones, soothing music, or a variety of natural sounds. After some experience, you may find your preference tends toward the more subtle sound of silence. The meditation centre you attend may favour a particular methodology. Brahma Kumaris recommended the open-eyed, sitting meditation method of Raja Yoga, and decades later, i still find this to be my preference. Closed-eyes have the advantage of eliminating one potential source of distraction – visual stimuli – though on the other hand, it is easier to fall asleep! One must be adaptable to the circumstances. For example, in a public setting, open-eyed meditation might be construed by others as staring, so to avoid their discomfort, use downcast eyes, even if that might be interpreted as sleeping!

For effective meditation, the attitude and degree of authenticity that one brings to it will affect how it progresses – or not. A goal-driven or impatient attitude will surely scuttle the attempt. Better there be calm feelings of relish, joyful anticipation, and surrendering – with no expectations. Meditation is a neutral process, and available for all – religious believers and non-believers alike. One should not try to "manage" the outcome of the meditation to conform to prior expectations, indoctrination, or other prejudices, but witness it with openness and sincerity. Let go of any prejudices and self-conscious-driven personae – the masks and dramas created by ego – and come into meditation unconditionally, with a genuine openness to witness; then we might reveal our authentic self to *ourselves* and the process. The authentic self has nothing to hide and need not be defensive, so the mind can relax from feeling it has to project being someone it is not. The meditation itself can help us find our true self, so that revelation does not have to be completed before meditating. Just be aware of the value of authenticity and drop the more obvious inauthentic masks ahead of time.

The act of meditation and staying in the moment is a fragile one. The meditative state can be susceptible to external influences, although the most frequent interruptions and diversions are those of the mind. These are potentially ubiquitous, penetrating, and provide a limitless resource for disturbing the meditative process. I find the best preparatory mental state is to become relaxed at the outset into a state of "soul-consciousness". The term "soul-consciousness" describes a self-induced mode of being, a philosophical frame of reference for the mind, which considers all from the viewpoint of primarily being a soul – or spirit or psyche – temporarily attached to this body.

That perspective is not exclusively for meditation alone; it can support an entire lifestyle of a soul-conscious being, in which the soul, spirit or psyche is the driving force of the body. Spirit or psyche is the root entity of the individual sentient being, which has a temporary body – like wearing a suit of biodegradable clothes! From a soul-conscious perspective, when using an expression such as "me" or "myself," the "me" or "myself" would refer to the soul, spirit, or psyche – the real me. The body occupied by me is just an impermanent appendage, a means of manifestation, interaction, and navigating this world.

This perspective of soul-consciousness provides different and often insightful viewpoints on events. One recognizes the impermanent nature of existence and relationships and realizes that all sentient beings are one family of primarily spiritual entities, each having a temporarily appended physical body. Spirit represents fundamental purity, only superficially sullied by the detritus of contemporary physical conditioning. It is a node within a distributed network of communal spiritual presence and purpose. To focus one's life on interacting with the world from the perspective of being first a peaceful spiritual being is to savour unconditional compassion for fellow beings, for all sentient and non-sentient aspects of our universe, to transcend material imperatives and figuratively float peaceably above the grasp of current woes.

Quieting the mind

A central challenge of meditation is to quieten the conscious monkey-mind from what Buddhists call "conceptual proliferation" – the tendency to leap frenetically from thought to thought. Quieting mental noise is not a battle; it is a subtle interplay between discursive thoughts, awareness, and intent, and not all monkey-mind interruptions to the meditation process are managed successfully. At first, the ego frequently inveigles itself into our conscious space and attempts to monopolize our quiet time. By adopting the perspective of soul-consciousness, we discover that humility and noisy ego are mutually exclusive.

One needs to be aware of any distractive thoughts arising, deflect them, and re-centre the mind. Do not attempt to inflict silence on the mind with rigid self-discipline. Gently and repeatedly, shepherd the mind away from its unlimited supply of self-made distractions, and toward the focus of the meditation. Initially, a focal point can be visual, such as some form of an external pattern or image, though i find a mandala's ornateness can divert the mind into pattern searching. It can be auditory, such as the tranquil sounds mentioned earlier, a mantra focusing on a particular virtue (a Brahma Kumaris favourite), or silent counting within the mind. It can also be sensory, such as relaxing the body one limb, zone, or chakra-centre at a time, starting at the toes, and working up to the crown of the head, or by maintaining awareness of the breath – in particular, the out-breath (a Buddhist favourite).

A beneficial attitude is one of "letting go" – of detaching from the myriad of worldly thoughts, concerns, and self-identities that otherwise occupy one's day. Let go of tension, expectations for the process, feelings of self-importance or self-denigration – being unpretentious – and be open to whatever comes out of the meditation process. If it appears to be nothing – that is fine, that is why it is called a practice. If there is some revelation, then that is fine too. Just remain in the "Now," having present-time consciousness, with no thoughts of past or future. Zen refers to this Now state as "being-time."

A notable meditation technique is the repeated self-questioning of one's identity. The contemplation of "Who am i" is a large part of achieving authenticity and loosening the grip of ego. Go deep; be persistent. The metaphorical image of peeling layers from an onion might be helpful. The onion is total self, and each "Who-am-i" question answered peels away another layer, revealing a purer core self. There is always another layer to peel away. We may not find the centre of the onion that is pure self, but that does not matter. Our task, by continually asking the question "Who am i," is to get past superficial self-identification such as name, status, vocation, or occupation, beyond physical identities such as gender, body, family function, national, social, or ethnic origins, beyond societally imposed roles and expectations, and toward the core of our existence. Determine who is the one asking the question – who is the one at the centre.

The one at the centre is the animating energy; the psyche, spirit, or soul, usually hidden from consciousness. Howsoever we prefer to label it; the one who meditates is the witness, the non-physical presence that co-exists with and relates to our mind and body. We can leave to one side any self-esteem issues, and remind ourselves that we are that pure, perfect presence, looking into the physical world through the eyes of our temporary body. Meditation helps us to detach from the mundane, egotistical aspects of the world and to become more aware of the true nature – spiritual, metaphysical, and material – of the phenomenon we call existence.

Frequently, we may need to remind ourselves to apply no effort of will, no judgment, no expectations, no analysis, and no discursive thought during meditation, but just to gently quiet the mind. Discourage conscious acknowledgement of being in meditation, recognition of what may be happening, or any self-monitoring of performance or reactions, just lightly quiet the mind. If we cannot pre-determine what kind of meditation we want to do, default to a basic receptive type – one where we have no agenda and are open to just going with the flow as a witness. There will always be the possibility of an excursion into mindful "Who am i" questioning.

During a workshop-style meditative enquiry, one may realize that certain interesting information is arising, which needs conserving for further contemplation. A head-on interrogation of this information is unlikely to succeed and can collapse the meditation. Instead, a respectful kind of indirect awareness may be more effective, rather like obliquely viewing something through peripheral vision. This mental approach indirectly observes inner thoughts, sensing and memorizing information without using the direct interrogative stare of the mind's typical focus. Thus, we can take information away from meditation without collapsing the process by avoiding active analysis and appraisal, or even the acknowledgment of it during the meditation; just covertly observing and memorizing while studiously avoiding conscious recognition of the experience. After meditation, the remembered information can then be processed while in contemplation.

Processing

In this materialistic world of treasuring certainty, permanence, and duality, exploring the spiritual path with any sense of rationality seems a minefield of ambiguity and paradox. Tolerance for ambiguity and paradox mitigates the frustrations of dealing with the many apparent contradictions and non-dual uncertainties that can occupy a spiritual worldview. Although a paradox may often result from limited powers of comprehension, it may also be due to a lack of appropriate information or perspective, such as might be encountered with an established dualistic mindset.

The inability to see something from the differing viewpoint of another; intolerance toward the non-conformity of others; the rush to judge or characterize; the inability to simultaneously comprehend two or more seemingly opposing points of view – these are all indicators of an immature intolerance for ambiguity. Part of formal Buddhist training is to improve that tolerance. One of the purposes of the famed ambiguous and paradoxical Zen koans, those riddle-like rhetorical puzzles, is to tease and tire the student's chatterbox intellect into dropping attachment to pre-conceived notions about self and reality. Typically, they will consist of a seemingly unanswerable question, paradox, or ambiguous phrase that makes no apparent sense. The student will contemplate the koan over lengthy periods, trying to find a solution or an understanding. Often, there is no pre-determined right answer or correct form of response – a solution may not even exist. Eventually, the student becomes so intellectually tired that he opens to tolerating apparent ambiguities, which may then result in a profound revelation – the intent of the koan.

As we start to experience meditative insights – as fragile nuances filter their way into our dormant consciousness – most revelations may make little sense initially. One should refrain from rejecting them because of their seeming lack of clarity or relevance. During the meditation, one should refrain from reacting at all – merely witness. Tolerance for ambiguity can prevent any proverbial babies of enlightenment from being thrown out with the turbid bathwaters of confusion. Instead, during subsequent phases of contemplation, one can probe mindfully around any ambiguous issue, review it from different vantage points, and reinterpret. Ultimately though, if still without resolution, then with all humility, allow a paradox to remain a paradox; do not discard it, it may jump into focus later.

The processing of any content resulting from the meditation is best started soon afterwards while still in a meditative or contemplative frame of mind, to retain the freshness and authenticity of the original experience, thereby minimizing the potential for imposing the rationalizing filters of active left-brain consciousness. Contemplative post-processing allows the information recalled from the meditation to be respectfully cogitated and interpreted, perhaps becoming the basis for further meditative exploration or even external communication. Contemplation allows the reflective mind to ruminate responsibly – to circle wider and follow more links than might be prudent during formal meditation. It can be a time for performing thought experiments to test the practicality of understandings gleaned from meditation. Although a subjective process, this is somewhat similar to that employed in scientific enquiry.

We have become so used to thinking symbolically – relying on the immediate unconscious conversion of observed experience into an encapsulated concept – that it is unusual to have to remember raw unprocessed information. While in a productive meditation, we are essentially dipping into unconditioned information with emotional overtones, while obliquely trying to retain memories of that experience. The process of moving unconditioned information in and out of memory for subsequent conceptualization is an unfamiliar one. Perhaps this explains the air of mysticism that often shrouds the reporting of such experiences.

Rendering

Rendering a meditative experience involves interpreting the acquired raw information that swims into one's awareness. Emotional elements may predominate rather than clear images, and post-meditation rendering converts the innate information into a form the conscious mind can more easily digest. The rendering process will necessarily use analogy and metaphor, so the larger one's knowledge base and conceptual skills are in that area, the more fluent the interpretation. As an evolving species, we should be better able to render and communicate meditative experiences than our ancestors would have been. If we were not so distracted by lingering remnants of ancient dogma and by our fascination for shiny materialism, such interpretive wisdom might be the basis of new metaphysical science. Indeed, some Buddhists would argue that this is how they treat contemporary soft-science studies based on their traditional understandings and philosophies.

The process of post-meditative rendering is delicate, requiring us to probe fragile memories of an ethereal experience. Treating these recollections with respect avoids projecting prejudicial views onto that frail information. As the process of disentangling the remembered experiences occurs during contemplation, frequent recall of those memories allows us to compare them with the developing conscious renderings to verify their accuracy. Like an artist, attention focuses on the process of building up the image, while referring back to the memory of the original subject to verify the image's accuracy and to acquire omitted detail. Through this rendering process, a clearer picture or meaning may gradually form. Sometimes, though, it can snap into place as a complete understanding.

Perhaps recognizing the limitations of rendering insights into the written word, the Buddha is reputed to have insisted that his teachings not be documented. Even so, many of his teachings survive today, though he was not the one to write them. He was probably concerned that widespread untutored distribution might become prescriptive.

Generally, it may be better to refrain from discussion of personal mystical illuminations. Even if sharing one-on-one with a like-minded peer, each party can come away with different understandings of the other's described inner experiences – unsurprisingly perhaps, since the experience is entirely subjective, as is the telling and hearing of it. Nevertheless, despite the potential for distortion, there does appear to be some conformity of understandings arising from expressed contemplative realizations expounded over generations. Called the Perennial Philosophy, these describe the universal recurrence of philosophical insights, independent of epoch or culture, and including realizations about the nature of reality and consciousness.

If sharing the content of one's meditation with an esteemed other, adopt the same respectful attitude for the subject as in contemplation. People generally like to stick to their beliefs in the absence of overwhelming evidence to the contrary – and frequently despite such evidence – so do not expect full accord with such shared content. Nevertheless, if ever a secular metaphysical science of subjective enquiry were to be established, the cumulative garnering and sharing of such acquired knowledge could become the catalyst for propelling humanity's psycho-spiritual and metaphysical understandings of existence into evolved levels of comprehension and clarity. Prudent sharing of such experiences may then benefit not just humanity, but collaterally, all beings.

AWAKENING MIND

If ever the practice of meditation were to be ascribed with but one objective, "awakening" would be it. Below are some general descriptions of the awakening state with a few references to associative states recognized in different cultures, followed by some samples of my own experiences. These immanent experiences are intimate, and the descriptions wholly inadequate. They represent a meagre attempt to describe the indescribable – to capture the wind with one's fingers.

Subsequently, when pursuing background research for developing this book, i found many traditional descriptions of these kinds of experiences. Although some descriptions appeared in the West quite recently, schools of Buddhism and Hinduism have been identifying these states for thousands of years. Thus, from this experiment of one, i have concluded that not only are such experiences not unique; they are available to anyone open to them, seekers and non-seekers alike. In other words, it seems unnecessary to search the literature first and become knowledgeable in this field to experience various forms of awakening. However, it does seem reasonable to expect there might be a higher incidence of recognition of such experiences among those who have already become conversant with formal meditation, all the better to recognize and accept any awakening experiences.

Terminology

Many esoteric-sounding terms are associated with realizing an enhanced state of awareness: words such as peak experience, awakening, enlightenment, *satori*, *samadhi*, and *nirvana*. Though these and many others can be a confusing array of traditional terms at the outset, it soon becomes apparent that the meanings of most converge. The most recent term, "peak experience," is perhaps the most accessible. It is scientifically connected, although it describes a similar state as the more traditional expressions. When the meanings of those traditional terms are examined, a common theme becomes evident, even though it may be described by reference to different metaphors and language, and the profundity and longevity of the state may differ. I use mainly Buddhist expressions here, in deference to that tradition having amassed a more substantial experience base than any western culture. Including the pre-Buddha period of Indian mysticism, it spans a period of some three millennia of record. Most expressions are of Sanskrit or Pali origin, although some originate in other Asian philosophies.

Even within the traditional Buddhist schools of Theravada, Vajrayana, Mahayana and its subset, Zen, the same terms can have different meanings. There is no international standard of definition, no common scale of experience for reference and measurement. From my limited experience, it seems likely that there is a continuum of transcendental understandings and states of mind, already available to each of us for discovery or recognition. The different labels merely identify by how much, and for how long practitioners have lifted the obscuring veil, glimpsed, and understood. The summarized descriptions of these terms included here are intended as an informal introduction for an aspiring meditator, not as a reference.

The psychologist Abraham Maslow, of Maslow's Hierarchy fame, coined the contemporary term "peak experience" in the 1960s. He described it as *"certain transpersonal and ecstatic states, particularly associated with feelings of euphoria, harmony, and interconnectedness. Such experiences and associated revelations have an ineffably mystical quality or essence. Characterized as involving sudden feelings of intense joy and awe, they may also include awareness of transcendental unity or knowledge of higher truth, often revealing vastly profound and awe-inspiring perspectives on, and affirming the value of existence. They tend to be uplifting experiences, transcending ego, and often all sense of self dissolves into an awareness of limitless unity."* … Right on!

Within traditional descriptions, awakening is the objective, and that term is commonly interchangeable with the word enlightenment, though awakening might be a more apt metaphor. Extended periods of awakening – known as *bodhi* in Buddhism – largely correlate with that of *nirvana* in most Buddhist schools as being a state of becoming immune to greed, hate, and delusion – the primary forms of suffering and sorrow. Described variously as complete and perfect sanity, or awareness of the true nature of the universe, awakening implies the transition toward a state of universal consciousness, rather than an end-state of omniscience. The state of awakening, enlightenment, or awareness can seem as if waking from daze or delusion – becoming aware of and knowing the underlying reality. Implied in this description is a realization of non-dual existence – eliminating the split between subject and perceived object – and the essential emptiness or absence of inherent presence in phenomena and physical things. Extreme attributes to this state include approaching omniscience, knowing things as they are, and having inconceivable knowledge. However, most sages agree that the only possible way to comprehend the state is through experiencing it.

Buddhists generally consider *nirvana* as the unconditioned mind, which has come to the point of perfect lucidity and clarity. It is a calm enduring state of transcendental contentment. *Nirvana* is not a location or end-state such as the religious construct of heaven or paradise but is an advanced state of mind considered by Buddhists as a person's highest spiritual attainment. To the extent that *nirvana* is attained, it becomes a ground of being from which to act ethically and holistically. The Buddha himself described it as *"consciousness without feature* (not conditioned by the senses), *without end* (limitless, uncontained*), and luminous all round* (without notions of duality, but open and uncompromised)." The parenthetic comments are mine.

While discussing *nirvana*, we should perhaps mention *samsara*. *Samsara* is a Sanskrit word equating to the cycle of life, death, and possible rebirth. An ancient and surprisingly cynical interpretation of it is *"that stage of wandering and suffering between birth and death."* The chief point is that, unlike *nirvana*, this state is not one we are endeavouring to attain since we appear to have thoroughly achieved it already. *Samsara* is the apparent material reality of the physical world of confusion and delusion. Its existence is characterized by suffering, which mainly results from ignorance. Many consider *samsara* as the opposite of *nirvana*. However, this view is dualistic since it implies that to evolve one should shun *samsara* completely and pursue only *nirvana* – often interpreted as rejecting material wealth, pleasure, and "earthly things" and becoming a monk. However, during the second century, the Indian Buddhist philosopher Nagarjuna expounded the non-dualistic view that *"samsara and nirvana are not-two."* In other words, they are not opposites, but are complementary and inseparable, like sides of the same coin. Only one's perspective needs to change to evolve toward incrementally heightening the state of *nirvana*, acknowledging *samsara*, and transcending beyond as well as including it – necessarily accepting the whole coin.

While it may not be clear what differentiates these various terms, it is apparent that many are largely synonymous, with only subtle distinctions of transition and degree. Two other terms illustrate this convergence of meaning – s*amadhi* and s*atori*. *Samadhi* is a Sanskrit word used in both Hindu and Buddhist philosophies, denoting the higher levels of concentrative meditation. Defined as a non-dualistic state of consciousness, it is one in which the practitioner's mind becomes one or merges with the experienced moment, during which the mind becomes still, concentrated or "one-pointed." *Samadhi* includes the initial experience of awakening – reaching toward a state of complete unity – and marks the beginning of the process of attaining self-actualization.

The earliest of the four stages of *samadhi* is a state of joy, profound well-being, and a peaceful, meditative state of mind. The next is a temporary state of full s*amadhi*, in which the mind becomes quiescent and gives up its desires and busyness. Here, a taste of bliss and awareness of being – or beingness – occurs, though some worldly attachment remains. The third level is the highest transcendental state of consciousness – complete unity – where there is no duality (no subject-object relationship). At this level, the differences and attachments of normal-world reality have faded, and all appears as holistic and complete unity; only pure awareness remains. Entering and holding onto s*amadhi* takes a concentrative meditation effort – meaning the mental managing of mind and ego, not fierce willpower. The first three stages are temporary, so eventually one returns to ego-consciousness. However, the fourth stage is the ability to remain in full-on *samadhi* and yet be fully functional in the perceptible world. Only the truly enlightened can stay in such a continuous state, but even individuals with no spiritual preparation or disposition can experience the initial stages.

Satori is a term used in Zen and is close in meaning to enlightenment. It is an understanding, an emergent transcendental awareness, wisdom, and individual comprehension. Considered the early stage on the path of *nirvana*, *satori* is an experience that can lead toward full enlightenment – a state of freedom from suffering, desire, and ignorance. Some practitioners believe that when experiencing *satori*, there is a psychological change that heals a perceptive blind spot in one's notion of reality, thus enabling one to be aware of, mitigate attachment to, and see beyond the delusion of duality. In traditional Zen, one attains *satori* through meditative effort, personal experience, and the use of koans. *Satori* refers to a profound and lasting clear realization of the nature of existence and a remembering of who we are. It has been described as the *raison d'être* of Zen.

Personal Observations

Despite their differences, these traditional terms all point toward a state of higher consciousness, one not easily described even by those who have experienced it, let alone understood by someone who has not. Perhaps attempting to add personal descriptions here is futile; nevertheless, i will still share some early experiences of these altered states of consciousness. I sincerely hope that my attempts to document these intimate instances do not result in a disservice to the experiences of others. These observations, which were noted shortly after the experience, are accounts of the mind employing metaphor in its effort to engage with something unfamiliar. Several of these experiences later served directionally to inform some of the avenues of speculative science pursued in Parts 2 and 3. Necessarily, most of these observations relate to examples resulting from sessions where a theme of sorts emerged, since it is hard to describe the no-image condition of encountering nothing. Nevertheless, some notes are included about sittings that did remain relatively neutral, and at least attempt to describe the post-experience emotional memories of having found oneself on the edge of nothing.

Before describing these experiences, i will mention an odd recurring emotive theme noted during several earlier meditation practices. I had been pursuing my quest and making some progress in contemplative meditation for some years when i started to get the feeling that a great revelation was hovering just beyond my grasp. The imagery accompanying this understanding was a familiar metaphor along the lines of hiking up a mountain but always finding yet another, higher peak beyond, challenging my ascent. Over several similar sessions, the feelings of the imminence of some revelation became more intense. Then the imagery changed to imply that after this mountain would be not the sight of a higher one, but of being at the top of a cliff-face. Once achieved, it would reveal a complete vista of the land below with unrestricted panoramic views of everything out to the horizon. However, i had to reach that cliff-top, then fly off it like an eagle, or else retreat down the mountain. It seemed to be some form of promissory awakening. No fear accompanied the cliff flight image, so i took it to be a positive incentive to continue the contemplative meditation regimen with relish, and even some sense of urgency.

Concurrently, there was a feeling that once launched clear of the top of that last mountain the reality of the world of mountains would become lost to me. Alternatively, perhaps, i would be lost to that world. It was not that i was going to die; it was not foreboding like that. It felt like a bland consequential statement of fact; the association between me and the reality i knew would cease, and i would be lost to it. What could that mean? I talked of this motif with my meditation leader, but she could offer no wisdom on the matter. The urge to see those uninterrupted views from beyond that last mountain proved stronger than any trepidation of the ceasing-to-exist consequence, which, at the time, i put down to a spurious effect of conflicted motives introduced by ego.

Many years later, i came to consider that there is a sense in which, after a peak experience of awareness, one does become lost to the old materialistic world of ignorance, because once aware, it is impossible to go back. Like the Buddhist doctrine of non-self and the Brahma Kumaris notion of soul-consciousness, it was perhaps an insight into a coming shift from identifying myself as a physical body, status, etc., toward identifying with the intangible psyche or spirit, the true me. If one's existence is defined by the degree of attachment to an illusory self-image and a materialistic world, then perhaps awareness of that illusion, represented by the flight from that last mountain-top, would cause it to cease existing for me – and with it, some of the old me. If this were the case, ceasing to exist in the former materialist world perspective was not a warning so much as a presaging of impending renunciation.

A Physical Peak Experience

Many can relate to a physically initiated peak experience, whether resulting from exertion, risk-taking, or performance excellence. It may take the form of detaching from one's investment in one's own body and observing oneself with unemotional objectivity as a remote and separate third party. Typical of such phenomena is the "athlete's high," which overrides any pain – a sensory-driven form of peak experience. Other stressful or emotionally charged situations can also trigger types of peak experience, such as joy, orgasm, exposure to mortal danger, traumatic incident, and, in the extreme, a "near-death experience." Even less severe events, such as breathtaking panoramic views, or an exhilarating ski run, can trigger an emotional wave that may lead to a brief peak experience of detached witnessing.

One purely physical activity that triggered peak experience symptoms in me occurred years before i started meditating. It was my first skydiving experience. After completing some ground training, the previous week, the plan was simple: go up in a small airplane to 12,500 feet above the ground and jump out, free-fall for about 30 seconds, then at 5,000 feet above the ground, pull the ripcord and float the rest of the way down under a steerable parachute.

"Several of us are jumping today. After take-off, the plane slowly reaches jumping altitude. Once there, since i am nearest the door, i exit the plane first. My fall quickly reaches terminal velocity, which i seem to remember for a human body in a spread-eagled position, as being around 200 km/h. Although this is quite fast, the ground seems to stay where it is, like a gently rotating photograph. Falling through the air at this speed creates a powerful wind that continuously pummels the body and deafens the ears.

Thirty seconds or so is not much time. I discover that by moving various limbs slightly, i can cause the body to yaw, pitch, and roll. That experience is exhilarating, and i take great delight in flying it – even accidentally doing a barrel roll. Note the mental syntax: the "it" here is the body – an "it;" no longer me, or mine. I feel i am just a remote observer witnessing the show, not aboard the body, but floating somewhere above the action, following my performance below, like playing a video game. The whole experience seems to be remotely unfolding before my mind, as a movie does for a theatre audience. However, even while actively engaged in aerial acrobatics, you are still descending extremely fast. Disappointingly quickly, as time runs out, the picture of the ground becomes bigger, noticeably faster. The dispassionate observer in me notes the rapidly unwinding altimeter, strapped to my chest, has just passed 5,000 feet.

Likely the most salient aspect of skydiving is that from the second you jump from the airplane; you have irrevocably surrendered to the exhilarating Now. You are committed to living in it and taking full responsibility for yourself and your actions. As in one-pointed meditation, the conscious mind is wholly focusing on the Now. No monkey-mind chatter, for the mind is quiet, save for formulating the imperative command for the body to do the next self-saving action. Remarkably reluctantly under the circumstances, i pull the ripcord. A disorienting jerk later, my chin hits my chest and i mentally reacquire my body to find myself suspended below an inflated, multi-coloured canopy. Another five- or ten-minutes pass, while i silently admire the ascending scenic views, hear the approaching barks of dogs, songs of birds, and the mechanical sounds of man − before the ground comes up and hits my feet. Adrenaline still pumping, i gather up the parachute and hasten back to the hangar. I want to go again!"

While an exciting experience, i am not suggesting that this kind of physically impelled experience can parallel the profundity of an awakening meditative experience. Nevertheless, some of the initial symptoms, such as detachment of the mind from the body, one-pointed concentration, and an incandescent awareness of our only reality being Now carry similar hallmarks to the beginning of a psycho-spiritual peak experience.

Mental Peak Experiences

Turning now toward calmer, mentally initiated psycho-spiritual occurrences, these are some lightly edited personal notes on meditative experiences, spanning several years of practice. Not every experience occurred during formal meditation; some occurred in periods of contemplation, relaxation, or dreaming. None of these insights provides a clear total picture; they are more like breadcrumbs, which can lead toward awareness of an alternative worldview. The first three resulted from neutral or receptive style meditations:

> "On emerging from the experience of being one with all, i, my heart, my body, are filled, overwhelmed, enveloped, and cocooned, with such a profound feeling of pure love – unconditional, undirected, uncomplicated love; accompanied by warm, blissful, serene feelings of belonging, security, home, finality. Then mortal consciousness takes over again, prying this experience away from the reluctantly fleeing "now" state of the unconscious mind, and the conscious mental analysis and trickery return.
>
> My left-brained consciousness cannot deal with the remnants of this unconditioned pure love; it seems to need to compromise that felt-love, direct it, source it, assign it, label it, and unwittingly conflict it. In the material mind, it could become labelled as an experience conditional on being "good," "holy," "chosen," "deserving," "giving," or even "meditating." It can become directed towards a personal or religion-owned god, its agent, a society, a group, or a person. The state can become offered by others as a reward for obedience, prayer, conformity, favour, and bizarrely, offered in trade for love itself – and it is lost.

These temptations of the conflicted, material mind need to be suspended, dismissed, or rendered neutral. For, to compromise this experience, which can only be described as totally encompassing love because no other words come close, is to corrupt a sacred communion with that holistic existence, that all-encompassing environment, which transcends the meaning of words such as love, compassion, reverence. That pure, unconditioned love is all that there is. There is nothing else, nowhere else, no when else, no other time than now. It is the essence of the undifferentiated consciousness in which we and our universe exist and complete the physical experience."

Another example:

"The first impression is a state of awareness that one seems to fall, rush, dive or float into, or arrive at. The emotional sense is like a sports-related "personal best" adrenaline rush, followed by warm feelings of detached wonder and awe. There is an excitement of a dawning understanding, recognition of what one is witnessing; and confidence that this knowledge – this newfound moment of lucidity – is both meaningful and sure. In Buddhist terms, i think this may be the stage called samadhi, or the Gate-less Gate because it seems you discovered something new, although it was there all along – only your awareness of it has just opened. You think you have located the hidden gate leading into awakening, though once you pass through, you realize there is not, and never was any gate, just an opening; awakening was always already available. Although the realization is that there is no barrier to this perceptive state of mind, there still seems to be a threshold over which one must pass for it to be revealed or recognized.

> Accompanying this realization is a sense of oneness with all; a lack of opposites, non-duality, a resonance, a linkage with everything, and every non-thing, a recognition of the essential, almost tangible, characteristic of being the Universe. The Buddhist term "One Taste" comes to mind, and now makes sense, because you embrace and feel embraced by all, one thing is not more important, more valued than another: nor less, nor even distinguishable. You are the silent witness now, for you have left any attachment to your body or your world and are part of everything and everything is part of you. Who remains to witness; no one, yet all. You enfold all and are enveloped by all. One experiences a profound sense of compassion and love for all sentient beings and respect for the non-sentient."

These meditative experiences were quite an emotional rush. The experience felt positive and desirable to repeat. It could become addictive if you failed to remain detached from the magnificence of the soaring feeling, and the instinctive grasping for the novelty of the revelation. However, letting ego charge in to ravish the occurrence would assuredly collapse the current session, and likely cause failure to stimulate another.

Finally, another post-neutral session expression:

> "Sometimes, very few times – around the level of experiencing the enfolding and being enfolded, enveloping, and being enveloped, merging as one with

all – another state or phase can occur. This next state is somewhat opposite to the image- and information-filled ones. Yet it is more profound. It is a calm state of just being. Ego is long gone, but now the observer, the remaining sense of self, that was part of previous experiences – that silent, choice-less witness – has also disappeared, dissipated, evaporated into the nothingness. Nothing and no one remain to observe – no observer, no observed, no subject and no object. Maybe this relates to the Buddhist term of "emptiness," i do not know.

Just a soft blackness remains, though not even as substantial as black would be. It is the absence of all sensory stimuli. From being aware of all, one has transitioned through all into this state of being one with none. The Zen term for this state may be satori, i believe. Many other words identify this condition, label it, and attempt to describe it. All fall short, for it is indescribable and only first-hand experience can awaken one's consciousness to it. Even then, one cannot completely grasp what it is; it seems more natural to describe it in terms of what it is not. If one brings conscious awareness to bear on the experience at the time, it disappears, and one is left with clumsy, inferior memories – ashes of a divine encounter – and a magnificent sense of humility."

The next few selections developed into image-driven themes from initially neutral meditations. They did not necessarily turn into "workshop" style themes, though. In subsequent contemplations, they invoked a mind's eye of unfolding visual images and vignettes of cognitive experience. The titles of these episodes i appended much later while composing this book.

God's Farewell

The following example was at an early stage in my meditation practice, and my perspective – even as an adult – was still influenced by my Christian schooling, and the more recent Brahma Kumaris philosophy, both of which posited an external, separate deity. In retrospect, this meditation was perhaps an example of beginning to let go of that stereotypical indoctrination, and transcending the externally imposed god-as-patriarch religious narrative:

"I am swimming, flying, drifting through a starless space, looking for God. In the distance, faintly, subtly, a pale bluish orb appears in view. I approach. Is this Earth, i wonder? Still, there are no stars, no sun, no moon, just this ghostly orb floating in velvet blackness. As i approach, i look around for God but do not see Him.

Now i am standing on a grassy slope overlooking a small, pale bluish lake. My perspective is as a little boy. I can feel God standing tall next to me, also looking at the lake, holding my hand as a father would. Out of the corner of my eye, i can see a sandaled bare foot the colour of marble, and the edge of a white robe beside me. With my other hand, on a childish impulse, i throw a pebble into the lake, sending a shockwave of ripples throughout the tranquil water, until the entire surface is disturbed. I sense an unspoken mild rebuke from God and wonder why. I realize that i am responsible for disrupting the serenity of the lake that we were contemplating, and which the lake was experiencing.

I turn my head to look up into His face, to see what depth of expression of disapproval might lie there.

> There is no face, no robe, no foot, and no hand. There is nothing. There is no grass, and there is no me. As i turn my gaze back to the lake, there is no lake. Only a ghostly pale blue orb floats in a black ocean of emptiness. And it is not Earth; it is the Universe."

Eversion

This next episode, although a separate meditation, seems to follow from the last theme:

> "Surrounded by comfortable soft velvety blackness, i perceive the Universe is drifting toward me, or i toward it. Its size increases until it fills my field of view. I am about to pierce the orb, to plunge into its being, to bathe in its essence. At that moment, my view distorts, as if when holding a powerful magnifying glass too close to the eye. The viewed image everts itself. The outer edge of the universe becomes the middle, and somehow i am now enveloping it. I both penetrate this orb and embrace it. I have entered it and swallowed it. It has surrounded me and lies within me. I am the Universe, and it is i. We are one. We are everything, everywhere, eternally. Immense feelings of satisfaction, joy, warmth, comfort, serenity, realization, release, belonging, freedom, knowledge, and certainty overtake me, it, us.
>
> A realization develops within me of nothing but comforting blackness and bliss. Then there is nothing, not blackness, not bliss, not even me; absolutely nothing at all – emptiness."

Bliss Remembered

This example continues the paradoxical theme of being small, encompassed by the universe and yet enveloping it. Perhaps a lesson in non-dual oneness:

"A gradual awareness of black emptiness occurs – a dawn scene unfolding to nothing. I am at one corner of space looking across at no stars, just empty, soft, black space. Yet i am not at the corner; i am in the middle, but i am not. I am not me. I am not at one point. I am everywhere. I am all. I love this nothingness; i belong to it. I wear it; it comforts me. I feel reciprocally loved by this nothing, this all. I give myself to it with infinite love and joy. It is i, not separate. It and i are one. We both are one; no other remains – no me, no i, no it.

Very gradually, awareness of a faint presence develops in the upper right quadrant of this experience of blackness. It is a vision, though not an image, an awareness of a soft unfolding rose. No colour is evident; just pale silvery-grey petals – delicate, velvety petals. On the scale at which i thought i was looking across space, this rose must be the size of multiple galaxies, or perhaps the universe. I sense that i can turn toward the rose. I am being drawn toward those soft petals. My perspective, my point of vision – there is no sense of i as a physical body – is that i am moving toward these petals until they fill my view. I am falling slowly, gently toward them as each begins to unfurl. Not falling but floating toward them at a slow pace.

The slow motion is like looking at the scenery far below through an airplane window. I am drifting into the petals, into the heart of the rose. As i approach each petal, it opens, unfurling to allow my passage and to reveal another one. This sense of falling, drifting through the centre of the rose, passing petal after petal, continues. Each one has a soft velvety texture like a jeweller's display cloth, yet i feel no physical contact. I am stationary. The rose is passing around me. I embrace the joy of fearless falling, of wondering without trepidation, and an aroused anticipation of where this journey might lead.

At last, i am through the rose, and all is black again, not even black, not anything; empty. I know, i understand without question, without doubt, i comprehend that i am back where i started. I love this place, this all, this nothing, so much, so deeply, with all my heart that it aches. And it is me."

These next two episodes seem more illuminated by technical metaphors and resulted partly from theme-guided meditation sessions.

Oneness

A sense of bi-directional association between love and creation is present, a hint of the paradox of being self-created, and intertwining of interior and exterior perspectives on existence.

"I am contemplating space, a vacuum. I am blackness, emptiness, an absence of anything, not in a dark sense, in a joyous sense. For, this absence contains everything. This nothing is all. This nothing is all that there is. I love and am loved by it. This is home to me, my origin, the infinitesimal point at which i start, and at which i finish. There is no in-between, for there is no start or finish, there is no time, there is nothing but love. That which i need – and i need nothing – is created by love. It is conceived, formed, and fashioned by love; it becomes love manifest. Within this emptiness, i can create anything with love. I can create worlds and universes, atoms, and beings, and live in them, be them, and let them roll around my tongue and taste them. They are inside me; i am outside them. I am inside them; they envelop me. I am them. We are one."

Regression

This second example appears to be a further evolution of earlier experiences, although it brought me into contact with the toroid concept. In this and many subsequent sessions, the toroid concept was to repeat and seems a significant metaphor. I treat this as a koan and struggle still to satisfactorily understand it and its potential geometric, extradimensional and scale implications for existence.

"I become aware of a ring, suspended in blackness; its size seems larger than the universe. As it becomes closer, more robust, it takes on a pale metallic silver-grey colour, with a dull shine. From my vantage point, which is nowhere, i can see into and through the centre of the ring. As i approach it, or it approaches me, i observe that technically it is not a ring; it is a torus, a toroid or donut, a long thin cylinder or bar that has been bent back on itself in a circle so that its two ends meet. It looks like a thick, round ring, with a slightly flattened cross-section, like a man's wedding band. My perspective leads me toward its virtual centre, and the ring is not much larger than i am. I can see that it is slowly rotating around its annular axis, like a perfect smoke ring. I think to myself; it would have to be rubber to do that, yet it seems to be more like a pliable metal.

As i pass into its centre, the sides of the ring that i can see appear to be moving, rotating at almost the same speed that i am passing through them, so the inner surface seems to be nearly stationary, relative to me. Now i am at the centre, the horizon of the ring that is behind me is evolving new surface. The horizon i am approaching, as i pass through the cylindrical part of the inside of the ring, is tantalizingly hiding the outer surface that i am just about to see. I seem forever poised at the very moment of seeing though never seeing what is on the far side of the disappearing surface of the ring. The appearance of the inside of the ring shifts from a short distance between the front and back sides of the ring into an elongated tunnel, through which i am travelling. The walls of this tunnel are moving with me, without my making much relative headway.

Time passes. The tunnel widens, and the walls turn back on themselves as they continue their curved journey around the outside of the ring's annular axis. My perspective expands and splits as i emerge from the back of the ring, so that part of me follows the tunnel walls as they become the outsides and then the front edge. I am enveloping the ring, orbiting the small diameter of this everting toroid. The other part of me continues to emerge from the ring tunnel to observe a completely new perspective on emptiness.

I realize that i am back where i started, the ring is again in front of me, although its scale has changed, and now it is much smaller than the universe, and its black backdrop is nothing. I pass over it and around it; i engulf it; i swallow the ring. Now i can direct my attention toward the new-found area of space where i emerged, and find a comforting, belonging, joyous awareness of being home again, and of knowing this empty place for the first time."

Less evident in this description, though noted in several similar episodes, was a shift in the scale of perspective. During the circling of the toroid or reacquiring the view of the universe, there was an accompanying sense that the size had changed by many orders of magnitude. Each version remained similar in the manner of a fractal, always appearing self-similar no matter the chosen magnification or viewpoint position along the dimension of scale.

Creation

This last example, an out-of-the-blue experience, started as a neutral, non-themed meditation that dramatically developed into unusually elaborate imagery. Though perhaps tinged with religious narrative, it turned out to be a theme that was to be repeated in subsequent contemplations, and which eventually catalyzed the speculative notion of a Binary Universe, discussed later in this book.

"White nothingness lifts like a fog. Across from me, over a small patch of water – a slow-moving stream or pond – a weather-beaten older man with straggly white hair, dressed in a long, off-white raincoat, is sitting on a log among some patchy grass, close to the pebbled beach at the water's edge. A driftwood campfire is quietly burning in a ring of stones near his feet. He leans forward slightly, holding two short, weathered planks flat together, intently staring at them. With a sudden burst of muscular energy that ripples through his sparse body and out through his arms, he violently pulls the planks apart and immediately twists them so that, while still separate, they are now at right angles to each other, like an x. An accompanying sharp crack sounds out, like a balloon bursting, and two bright little sparks materialize in the gap between the planks.

The twin sparks hover above the ground, waiting it seems, stationary in front of the man. He contemplates them for a while with an air of satisfaction and then lifts his eyes, then his head, turning until he is looking directly at me. He wants to ensure that i have seen what he has done. His black, fathomless eyes piercingly stare into mine, he thrusts his jaw forward and jerks his head up, to emphasize his unvoiced question "did you see – did you understand?" It was important to him that i witnessed his action. I have seen, though i have not understood…yet. The sparks remain, the scene dissolves."

PART 2

Science

*"All our science, measured against reality,
is primitive and childlike – and yet it is
the most precious thing we have."*

- Albert Einstein (1879 – 1955)

THE HUMAN CONDITION

"Body, Mind, and Spirit" is a familiar phrase – a cliché even – used in existential discussions of the human sentient being, both within and outside religious contexts. This trinity spans the material and nonmaterial nature of human and other sentient beings and describes the integrated self. The phrase groups the whole human condition, what we consider the complete sentient experience, into three parts, ranking them from the material to the ethereal, from the physical to the metaphysical. For this account, we will examine the body, as represented by the brain; the mind, by conscious and unconscious processes; and the spirit, by a mainly nonconscious part – the psyche.

The Brain

The human brain inhabits a precariously balanced skull, once described as "a bone box teetering atop the spine." The brain is essential to and physiologically maintained by the rest of the body. It is a focal part of the central nervous system, containing billions of neurons. The brain is the centre of neurological processing, and although generally considered as the source of mind, it may not be exclusively so. Its design is still evolving, and it is significantly different today from that of our recent ancestors. It now has more information to process.

The development of language, conceptualization, and skills over the generations continues to alter its anatomical structure.

The brain is not a single entity, but a "system" comprising three interrelated brain centres. Though interconnected, each of these brain centres has its specialties and provenance. These centres do not merge to form a homogenous facility; they remain and act as separate components. Any one of them can confirm, conflict, veto, support, or hijack the intent of another, depending on perceived environmental circumstances, or any pathology that might exist.

Cerebellum

At the base of the skull are found the cerebellum and brain stem (together constituting the reptilian brain); it is the oldest brain component in evolutionary terms. Its development started as reptiles first emerged onto land. That was over a half-billion years ago. Over the next quarter-billion years, its evolution essentially reached its full potential. Its function has changed little since then. It includes control over basic autonomic physiology, such as fine motor control, heart, lung, and gut functions, and archaic, involuntary, or instinctive psychological responses – including defence, aggression, ranking, mating, and eating. It appears to be a significant centre of basic raw emotions such as fear and anger. Its functions are rigid, obsessive, compulsive, ritualistic, and paranoid. It is inclined to repetitious behaviour, even when wrong, and is reluctant to learn anew. This brain centre never rests, even during sleep or meditation.

Limbic System

Next up is the limbic system, the paleo-mammalian or animal brain, which includes the amygdala – associated with emotional, unconscious memories – and the hippocampus, associated more with conscious memories. Its evolutionary life started when animals evolved away from the reptiles, a massive divergent shift from "cold-blooded" egg-laying reptiles to "warm-blooded" young-bearing placental mammals. There is no clear-cut timeline for this critical evolutionary step, but it took nearly a quarter-billion years to stabilize. The mammalian brain had to be more complicated than the reptilian; it had to take in more information, process a lot more "stuff." Therefore, it could not just expand from the rigid reptilian brain; it developed as a separate layer on top but could integrate information flow with the basic reptilian brain.

The addition of this mammalian brain developed into a highly effective animal control centre responsible for: nurturing; feed, fight, or flight survival decisions; sexual behaviour; memory; and the processing and buffering of emotions. Its principal priority is the animalistic seeking of pleasure and avoidance of pain. It evaluates that which seems either "good" or "bad" for it, and buffers emotionally traumatic experiences – sometimes even to the longer-term detriment of the health of mind and body. It also ascertains what gets its attention – saliency. It began to deal with the subtler processes of emotion, intuition, learning and applied intelligence.

A mere ten million years ago, from within this animal kingdom, the forest-dwelling gorilla and chimpanzee line of mammals emerged. Then, within the last several million years, one of the chimpanzee forms, seeking some evolutionary advantage, dropped to the ground, emerged from the treeline, and became bipedal. That was the dawn of the protohumans. Standing and walking upright would already have been a major evolutionary change, however, what it then enabled was pivotal. An upright posture allowed a better-balanced skull to enlarge behind the top of the spine, thus providing the potential for enlarged skulls housing bigger brains and expanding intelligence. This same balanced arrangement of the head on top of the spine would also allow for the repositioning of the larynx and the back palate of the mouth, enabling modulation of primitive grunts and eventually, combined with a desire to communicate, would evolve into articulate speech. Thus was language born.

Neocortex

Though naturally nomadic, tribes formed, numbers increased, life got complicated. More brainpower was required. Nature's solution was again not to expand an existing brain centre, but to develop rapidly additional layering of a third brain within an enlarging skull. This extra layering started a few million years ago and, after some false starts, led to the development of the species to which we belong. This final layer of the brain is the neocortex (the neo-mammalian brain) and is in all senses of the word, the superior brain centre. It is physically larger, being roughly two-thirds of the mass of the total human brain, comprises the left and right cerebral hemispheres, and occupies the top and forward sections of the skull, above the other two brain centres. The outer layers contain uniquely human neurons, providing higher cognitive functions, which allow us to think rationally, linearly, logically, and symbolically.

The two cerebral hemispheres connect via the Corpus Callosum, which acts as an information highway linking the two, allowing for the processing of information by either hemisphere or both. The division provides for the well-known, cross-connected left and right hemispheres – colloquially, the left and right brain. The left, being primarily responsible for the right side of the body in most people, tends to look after the more detailed, rational, linear, analytic-thought processing and language. The right hemisphere generally manages the left side of the body and mainly works with the big-picture, less-detail aspects – the more intuitive, emotional, abstract, or inspirational-creative skills. The combined neocortex handles rational processing, recursive thinking, and the analysis of and responses to the experience of being. The neocortex is not unique to humans, though in humans it has evolved massively and rapidly to be proportionately more than twice as large as that of any other mammal, all within the last two hundred millennia. Within the last fifty thousand years – an instant in geologic time – following the development of language, art, and abstract thought, it has become well-established, though not fully stabilized, since it continues to evolve.

Astonishingly, this latest period of rapid development covers just a few thousand human generations. We are indeed a most recent species – young and immature – and it often seems we do not quite have the hang of our neocortex yet. Where meditation and psycho-spiritual matters are concerned for self-contemplation and spiritual learning, the focal point of access to this domain, though favouring a "right-brain" portal, seems to share input through, or be influenced by, the limbic system. The result may be that every time we use the neocortex for conscious thought, the left hemisphere can inhibit the more subtle influences of the right and animal brain. When that access closes, we lose touch with the awareness that each of us is one with all. We lose sight of and appreciation for non-material components of existence – the psycho-spiritual and metaphysical possibilities – and we begin to lose touch with the internal language of our body, emotions, and spirit.

That might seem a heavy price to pay for the ability to invent planes, trains, and video games. However, if we were to allow ourselves to develop more thoughtfully and bilaterally, perhaps we need not pay that price. Maybe we could have mindful rational thought as well as intuition and universality. However, we are currently so giddy with the power of our "new" analytic mind that we mostly ignore our intuitive side. This new gift of nature's evolutionary process has brought to humankind self-awareness – both a blessing and a sometime-curse. In prideful arrogance, we even labelled ourselves "Homo sapiens," the *wise* human. This new human race may be smart, yes, but wise? Well, the jury is still out on that.

The Mind

Entrained within the human self-aware stream-of-consciousness are higher thought-processes such as intelligence, perception, conception, reasoning, feelings, emotions, memories, and intentions, which the mind manages under the joint control of the brain's three centres. Most conscious thought-processing takes place in the neocortex. However, some may unconsciously interface with and through the animal, and to a lesser extent, the reptilian brain centres, both of which can then influence the mind's conscious stream.

Some claim the mind is an emergent property of the physical brain – one where thought and memory are simply biochemical processes confined within and wholly arising out of the brain's physical structure. However, there do appear to be unexplained sources of non-sensory "inputs" to the conscious and unconscious process streams – enough at least to significantly alter the characteristics of those mental streams from how they might develop if they were completely compartmentalized within the brain. These additional potential inputs need including as part of a complete description of the mind. Although the stream of consciousness that forms the mind seems circumstantially resident in the brain, where most senses report, some indirect evidence suggests that an individual's intelligence, intuition, ideas, and memory may not exclusively reside there. The stream of consciousness can also be influenced by the individual unconscious, which may include nonlocal elements such as might be characterized by influences traditionally identified as psyche or spirit. The term "nonlocal," borrowed from the principle of quantum nonlocality in physics, refers to actions or events occurring beyond the apparent limitations of spacetime, sharing mutual information faster than light could travel between them. We will be discussing the principle of nonlocality in more detail later. Meanwhile, we can note that throughout this book, a reference to the nonlocal mind implies nonlocal influence on an unconscious part of the whole mind.

The left hemisphere is thought to push the self-awareness, self-identity concept into the consciousness, creating the "self and others" dichotomy. In contrast, the right brain has more panoramic, holistic perspectives. Furthermore, the right brain may be implicated in having subtle, non-sensory access to information sourced beyond the conscious senses, nonlocally. It may participate in some form of a shared unified field of unconditioned consciousness – an observation paralleling the communal "collective unconsciousness" conjectured by Carl Jung. This critical aspect we will explore later.

An individual may thus be in a permanent state of mental tension between two different or at least complementary ways of perceiving and conceiving their sentient existence – dualistic detail and holistic panorama, which in turn may be informed partially from nonlocal sources. Meditation can successfully access this comprehensive state by quieting the adverse effects generated by an active left hemisphere, thus allowing the right brain's functionality to strengthen, at least for the duration of the session. The right brain can act as the chief interface between the individual's conscious and unconscious mind; it includes the focus of the psyche and may tap into a distributed, collective pool of hidden consciousness.

Think of remarkable phenomena, such as instinct, intuition, and inspiration, the extreme feats of geniuses and savants, and that still small voice we call conscience. Such mental richness is unlikely to be produced entirely by the limited complexity of billions of neurons, or even the quadrillions of associated synaptic connections that comprise the physical brain, entombed within its small bone box. Without considering some form of nonlocality, the local physical brain could not have the capacity to account for much more than a fraction of the potential and performance of the healthy human mind. Thus, these phenomena may exemplify nonlocal influences on the unconscious – contributions extending beyond the physical brain.

When meditating, and during profound contemplation or self-enquiry, it is helpful, even necessary, to diminish the impact of the conscious self-aware mind's constant chatter. Most of this chatter arises mainly from the linear, recursive processes of the conscious mind – the stream of consciousness. Nonlocal contributions, such as intuition or inspiration, can be overwhelmed by this activity, and are harder to detect while the conscious mind is so active. In the case of meditation, it is often those nonlocal unconscious contributions, usually obscured by the mind's conscious noise, to which we want to connect.

Interestingly, most of the mind's conscious chatter is about the self. It may be more left-brain-dominated than right- and even though it is mostly a conscious stream, much of its content becomes habitually tuned out of the mind's day-to-day awareness, because it is so ubiquitous. Ironically then, we spend much of each day mentally talking to ourselves about ourselves without even listening to what we are saying! Much of this soliloquy relates to self-esteem issues. Humankind seems to have inherent self-esteem challenges – the innate need for self-justification and validation – and they centre on ego, an artifact of the self-aware mind. It may only be a mental figment, but it can be an intrusive one - one which threatens the effectiveness of meditation through distraction. That is why meditators emphasize the benefits of diminishing ego or adopting an egoless perspective.

States of Mind

The quality of self-awareness, contemplation, introspection, or self-enquiry experienced in meditation will largely depend on the nature of what else is at the forefront of the practitioner's conscious mind. When using these methods of exploring one's inner unconscious mind, the state of the conscious mind is of critical importance. On one hand, we must be alert enough to monitor our mind's behaviour. On the other, we must limit the conscious thinking activity sufficiently to avoid collapsing the entire subtle process. This balancing act of mind-states is, in many ways, the most challenging aspect of meditative practice. It is a skill made harder to acquire because the intense application of conscious effort to maintain that balance is likely to be counterproductive.

Various disciplines have come up with several terms to describe, differentiate, and further define the levels of alternative states of consciousness and unconsciousness in the human mind. Some of these, depending on which school one follows, may appear to conflict with or duplicate each other. Various prefixes to consciousness such as un-, non-, sub-, pre-, higher-, super-, uber-, and cosmic have been used in the attempt to distinguish between the various apparent facets or characteristics of the hidden mind.

For the terms used frequently in this book – consciousness, unconsciousness, and nonconsciousness – there appear to be no universally agreed-on definitions. Consciousness, and its derivatives, is just another one of those words that, the more you think about it, the more elusive becomes the meaning. However, perhaps we can outline some simplified definitions of these terms – particularly of the hidden parts of the mind – as they relate to the subject at hand.

Consciousness, as referred to here, is the alert, waking state of an individual mind that is aware of itself, its environment, others, and the relationships between them. Included in its repertoire are self-awareness, sensory processing, sentience, intelligence, and rational, conceptual mental processing or thought.

The balance of hidden mental processes tends to fall into the overarching term of unconsciousness. Active states of mind may still respond to external stimuli, but any cognitive and psychic processing occurs in the absence of self-aware consciousness. Within this meaning can be found many grades, extending from bodily autonomic reflexes and primary sensory processing, right through to dreams and absorbed spiritual awakeness. It embraces processes of the psyche or spirit.

Included in these processes are nonlocal contributions – considered here to be the realm of nonconscious. Profound meditation bridges the conscious, unconscious, and nonconscious psychic processes. Such a hybrid state points toward a superior condition of universal awareness. By accessing the cryptic nonconscious processes of mind through meditation, an expanded state of consciousness can result. Such an elevated mental state, rendering full universal awareness – sometimes referred to as "big mind" – suggests that unconscious mental processes transcend the conscious parts.

If we were to represent the whole mind by an iceberg floating in the sea – nine-tenths of which is underwater – then just the tip, visible above the waterline, would encompass what we call consciousness. Although a smaller segment, this part of the mind is prominently responsible for self-awareness. It also enables rational processing of sensory input, thoughts, concepts, and emotions, which are mostly supplied by the unconscious. It can apply itself toward improving the long-term survival and betterment of self and species. The terms mind and consciousness may often be used synonymously. However, these pages maintain that within the whole mass of this iceberg-like representation of total mind are found many parts, of which the most visible – conscious self-awareness – is but a minor one. Concealed below the waterline is found the larger mass of unconscious mental functions, ranging from usually obscured operational aspects such as autonomic maintenance functions, to higher-mind facilities such as nonconscious processes associated with the psyche or spirit.

This still-developing model is not the only way of looking at the structure of the mind, however. Some Eastern sages traditionally identify just four levels of mental state:

- Awake

- Dream-state (a concept shared by many aboriginal peoples throughout the world)

- Dreamless sleep

- Witness consciousness – *Turiya* in Sanskrit – is a state of constant awareness in which one remains aware of one's physical status but experiences a clear recognition of one's limitless metaphysical circumstances, of universal awareness, and an existential unity or *nirvana*

Buddhist psychology (*Vijnanavada*) provides a more detailed model of the mind, which subdivides the total mind into eight parts:

- Sight

- Sound

- Smell

- Taste

- Touch

- Thought

These six parts are the consciousness of the six senses – note the inclusion of thought as a sense. The last two are:

- Ego-consciousness

- "Basic" consciousness

A more descriptive phrase than "basic" might be "foundational" or "fundamental," since, despite the term "basic," it is vast – i imagine limitless – though we are not usually conscious of it. This level of basic consciousness appears closely allied to concepts espoused in the Tao, Jung's collective unconscious, and the filter theory of the psychologist, William James. It is considered fundamental – a storehouse of all potential – or as unity, and can be described as universal life, the reflection of which is found in our "higher mind unconsciousness." Historically, it was depicted as "a seed source for manifestation as form, phenomena or thought, acquiring the perfume of experience and then maturing into more seed." It is a germinating conscious-energy potential in a process analogous to biological life, where a seed grows, manifests its potential, incorporates experience, matures, and then begets more seed. In Nature, this is a life cycle, and when repeated, describes evolution.

Another term frequently used throughout this work and closely allied to consciousness is sentience. In Western philosophy, sentience can mean unconditioned consciousness and the ability to have sensations, experiences, or thoughts, described as "qualia." In Eastern philosophy, sentience is a metaphysical quality associated with all things capable of suffering.

Spirit, Psyche, and Soul

The term "spirit" is a long-lived traditional label for that vital though hidden animating force of all beings. It is the noun from which comes the adjective "spiritual" and is the third component of "Body, Mind, and Spirit." In a secular sense, spirit shares similar meanings with "psyche" or "soul." Psyche is a term used in psychoanalysis and other forms of depth psychology – indeed; it is the root of those words. Psyche refers to the life force or energy in an individual, influencing thought, behaviour, and personality. Although seeming a more contemporary term as compared with spirit, psyche is borrowed from the ancient Greek (*psukhe*) for soul, mind, breath, or life. It also means butterfly in their mythology.

There appear to be more similarities than differences between the terms spirit, psyche, and soul. All three words primarily attempt to describe the same timeless but ephemeral, non-physical but influential, non-egoistic but self-aware part of our entity. They are mere labels for an apparently intangible though vitally essential and foundational part of our existential make-up – a vital force, an inner essence within living beings, part of the actualization energy of an individual sentient self. As in multiple-witness accounts of the same incident, variations of description do occur. A trait attributed to spirit, psyche, and soul, is that it associates with the physical body for the duration of the body's life cycle, but upon death, ceases as an individuated entity.

Although the concepts of a temporarily attached psyche or spirit appear interchangeable here, in some religious interpretations the concept of a soul may imply a permanently recognizable part of an impermanent individual. That part is said to survive the body's death by remaining as a discernible character, which then goes on to enjoy the fruits of past religious observance in a new location – possibly heaven or paradise – in some non-corporeal though recognizable personified state. The egoic attraction of this interpretation is that it appears to extend the ego's existence beyond death.

The interpretation here is that the individual spirit, psyche, or soul is not emergent, not derived from the physical brain. It is fundamentally non-physical and nonlocal, and while not contained within the individual's skull, its centre of effect appears assigned to and focused through the mind and brain. Spirit acts through the nonconscious, which behaves as an interface with its source. During meditation, one can sometimes notice a physical tingling sensation just below the skin at the centre of the forehead, portending entry into spiritual communion. This active energy focal point is roughly in line with the pineal gland and corresponds to where some Asian cultures locate their religious *bindi*.

The entity described as spirit or psyche – or the non-egoistic version of soul we just discussed – associates with a temporary physical body. This combination is what we perceive as self. The primary forms of nonlocal spiritual communication with the conscious mind seem evidenced through the voices of conscience, intuition, instinct, and emotion, and during authentic meditation and possibly dreams. Spirit or psyche is partially accessible by meditation. It is the root and sponsor of our basic nonconscious driving force, and thence to consciousness and self-awareness – the spark of sentience within a sentient being. In concept, a person's spirit or psyche is but one of innumerable similar focal points, each perceived as a vital force while associating with an individual living sentient mind. Together, these form part of a universally distributed field of unconditioned consciousness. Every evolving awakened sentient mind can acquire unlimited access to the essence of this universal consciousness, subject only to the individual's capacity to absorb it.

The Sanskrit word *atman* means soul, spirit, psyche, and self. The ancient sages considered these as indistinguishable components of a sentient being. Other familiar-sounding labels applied to this intangible component of individuated sentience include: Life Force, Ch'i, Qi,

Tao, Prana, and Divine Essence. As noted earlier, within the context of this work, spirit, psyche, and soul, as well as such ancient terms, are relatively interchangeable in meaning. It was Spirit that led me to the findings described in this book though, which is what this section addresses; however, as the book's theme unfolds, i will often use the term psyche instead.

Physicalists — those believing consciousness emerges from the physical — might deny any such notion and even see consciousness as being nothing more than an emergent physical brain-based phenomenon — merely a behavioural response. I need not try to justify the existence of that entity which remains within each of us — aware when all else is quiet — that silent witness. Subjectively, it is clear such an entity exists, and our best chance of communicating with its source is through meditation.

During a profound meditation session, a point can occur when attachments to all mental symbols, images, thoughts, and most feelings disappear. What remains is nothing, except the awareness of nothing. Then the question is, "Who is aware?" Who witnesses the absence of anything at this point? The witness of the last, at the last, is that silent witness. It is the divine in Us. It is pure spirit, authentic self — not the false projection of self that ego has been assembling throughout our lives. The core link between our nonconsciousness and the universal field of innate consciousness is a ray of the divine. If we can sustain that magical point without conscious thought, we can reach the state where even the feeling of witnessing disappears. That divine ray melts into grand nothingness and becomes the absence of anything. Temporarily, oh so temporarily, we become one with nothing.

Conscience – Evidence of Spirit

"That still, small voice within" describes conscience. It appears to originate from within the mind although is more likely the influence of the individuated spirit. Conscience is a moral faculty, a "natural" or "inner" intelligence, a light beyond the conscious, rational mind. The importance of conscience here is that its existence, which is widely and even legally recognized, provides evidence of spirit. Most people have noticed on occasion that conscience does shine through from the unconscious, patiently and persistently making its non-materialistic, altruistic presence known to the conscious level of the mind. Sometimes defined as becoming aware of the self-referencing power of discrimination, it is a personal moral compass guiding between a moral right and a moral wrong. In this context, right has non-egocentric, altruistic, and holistically moral and ethical value – putting the benefit of another ahead of oneself – and wrong does not.

Conscience is the unreasoned, emotive "knowing," or "feeling" that an action contemplated by the conscious self, goes against, or conforms to higher moral principles. Conscience communicates this moral dissonance or resonance to our consciousness through the same nonconscious channel as intuition, via the language of emotion. It can signal through the pit of the stomach, with feelings of remorse if an amoral action has already completed. Alternatively, if still contemplating that prospective action, another stress centre may be triggered, such as a heavy heart or headache. Like intuition, conscience does seem to provide some direct, albeit subjective, evidence of a morally elevated source of awareness – the experience of being guided in moral judgment from beyond consciousness.

Frequently, conscience is thought of as flagging a negative or amoral act. However, it can also flag and reward a positive moral action, and does so by instilling subtle feelings of confirming joy, elation, and even elevated self-esteem or rectitude – when you know you have done the right thing. Conscience is another example of a communication process taking place within a hidden part of the mind, through the nonconscious psyche, and bursting forth into rational consciousness using the subtle, unspoken language of intuitive emotion. That emotion-based language can express feelings of remorse, worry, or physical discomfort; and yes, sometimes joy and elation. Such transactions are not intellectual, conceptual, logical, or rational; they are negotiations conducted in the raw, sacred language of Spirit.

Spirit on a Universal Scale

While discussing spirit, consider the related idea of spirit on a universal scale. The concept of a universally distributed field of consciousness or sentience is a central theme here. Such a universal spirit would be in continuous network-like dialogue with each unconscious sentient mind by accessing nonconscious processes. This understanding reflects the idea of the universally distributed nature of an undifferentiated conscious ground-of-being in which we find ourselves, and has similarities with the Buddhist concept of "basic" consciousness described earlier.

This concept of a distributed field of undifferentiated consciousness may have been what Carl Jung had in mind when he distinguished a "collective unconscious" from the personal unconscious particular to each human being. Jung described this collective unconscious as "a reservoir of the experiences of our species." However, such a definition may only seem "collective" from the perspective of multiple independent human consciousnesses. The view offered within this book of a unified, universally distributed, though individually accessible, undifferentiated consciousness, comprises a plurality of single-point focuses, correlating with each one of *all* sentient beings – not just the human ones. These multiple correlates then give the psychic appearance of being a mass of individual, scaled-down holographic versions of the universal entity, which, because they attach to separate bodies, seem to be different personal spirits. Publicly, Jung did not go further into realizing the more profound implications of his conceptual understanding of this extra-corporeal collective unconsciousness, possibly because his overriding goal was to develop a practical science for healing the mind.

It is certainly not a new understanding that some cosmic-scale energy, intelligence, or consciousness influences the universe. Such a universal-scale spirit or field-of-consciousness concept is seen as the prime mover and enabler in the inspiration, development, and evolution of the emergent physical universe and the existence of sentient life. Possibly, the bases of many of the old spiritual "truths" offered by the great religions would have had this common grounding initially. Only later would they have become anthropomorphized, politically differentiated, embellished, and ritualized. It is a key concept throughout this book, and we will be touching more on its nature, identity, and purpose, throughout.

Somewhere between the mind and spirit descriptions of an individual is the often-underestimated quality of emotion, and, as we shall see next, emotion may have an understated role in influencing our perception of the nature of existence.

INFLUENCERS OF MIND

Emotion

Emotion is an underestimated, poorly understood yet powerful influence in our lives and it tends to be a messy, complex one often studiously avoided and even denied by many who would prefer it not exist. Emotion can influence the quality of meditation. Furthermore, its pathways may support communication between the conscious "i" and nonlocal, psycho-spiritual unconscious processes.

Intense levels of emotion – whether negative such as those associated with fear or anger, or positive such as those of joy or compassion – can make the body act unconsciously and can even suppress conscious contributions. Those unconscious, intuitive alert calls to the mind – which often arouse conscious activity such as contemplating the wisdom of specific thoughts and actions or paying heed to inexplicable gut feelings relating to remote events – include a strong emotional component.

Emotional influences on the mind can be communicated to others consciously, such as through word, image, performance, and music, or at more subtle levels, unconsciously. These might include empathic sharing of emotions such as compassion, happiness, joy, fear, or grief. Such emotional sharing between two or more minds, particularly if occurring at the unconscious level, may not depend on any of the conventional senses for communicating. It may be in the form of an emotional resonance between minds, which then acts as the carrier for transferring or sharing the emotional values or qualities. Anecdotal evidence for this kind of sympathetic emotional resonance – mood sharing – between identical twins, lovers, or otherwise emotionally close individuals, even though at great distances apart, appear to show such communication as instantaneous, not limited to physical interactions and possibly not even by lightspeed. However, instances of such sharing of emotional content are mostly reported anecdotally and so do not readily yield to experimental testing protocols, thus remaining scientifically unpalatable. These kinds of sporadically reported events usually end up in the catch-all category of extrasensory perception, thus dismissing them from "serious" mainstream science.

In Western culture, the subject of emotion has a somewhat negative image, due in part to its frequent association with irrationality – the anti-social, extreme expressions of hard-to-control adverse emotional reactions and outbursts – and therefore the antithesis of reason. So, throwing the baby out with the bathwater, our left-brain-dominated society tends to label all emotions as weaknesses of character, if not downright liabilities. However, we can learn from quantum mechanics here: a subatomic particle is not fully describable without reference to both its particle-like properties and its wave-like characteristics – it is not one or the other, it is both. Similarly, the psychological nature of a complex sentient being, such as a human, cannot be wholly described without reference to both reason and emotion. It is not dualistically one or the other, but both. Emotion may even be symptomatic of a universal driving force for all sentient beings – a form of non-symbolic intercommunication independent of language, creed, and even species.

The physical sense of emotion often feels as if centred in the heart, gut, or solar plexus; science credits its arising as being generally manifested within the mind. Research points to the mammalian brain centre for many emotions, though the coarser reactive emotions such as fear and anger might correlate more with activity in the reptilian brain, occasionally bypassing further mental processing and directly producing physical reactions in the body. Some loftier emotions such as compassion and nurturing may be associated with mammalian brain area activity and channelled through the right brain. Even though different emotions can "light up" different parts of the brain in laboratory testing, it cannot be concluded that these point to the origins of those emotions, just localized areas reacting to them. As with other unconscious events, emotional manifestations may be induced locally, or by nonlocal access through the hidden gateway associated with spirit.

Emotion as Language

The nebulous and unfocused nature of emotion does not readily respond to reductionist analysis or even detailed verbal description. Nevertheless, one could envisage emotions as falling within an emotional continuum, analogous to the spectrum of light. With light, the entire range of visible light comprises individually labelled bands of specific light wavelengths or descriptions, which fall into bandwidths broadly identifiable as colours. We call these bandwidths by names such as "red" or "green," or technical descriptions measured by wavelength or frequency. Emotion, however, is a much less understood phenomenon than light. Although we might have some feel for its relative intensity, we cannot measure or define its energy as the equivalent of a wavelength. Cognitive scientists can map out emotion-correlated paths and associated activity centres using brain-scanning devices. However, that is not a measure of the emotion, just the brain's reaction to it – just as measuring the volume of a radio will reveal little about the workings of the receiver, the source of the music or the music itself.

With the light spectrum, if you were to combine all the colours of light, you get a neutral white, and if you remove them all, you get black. Imagine if that same idea applied to the spectrum of emotions. What might be the white light equivalent for emotion? There is no correct answer since this concept of an emotional continuum is somewhat contrived, though if there were, it might be "compassion." Some might vote for "love" or "loving-kindness" as representing a totality of the emotional spectrum, and, since this is a subjective exercise with no unique solution, they would not be wrong.

While the measure of emotion remains entirely subjective, we have attempted, through the ages, to label the "bandwidths" within the spectrum of emotions. Instead of red, green, and blue, for example, we have positive emotional bandwidth labels such as peace, love, joy, and many such benign uplifting states. Then there are the sombre bandwidths such as anger, fear, despair, and other dark emotional states. It may not be an easy task to distinguish qualitatively between adjacent emotions. Still, if you were to take some emotions such as bliss, joy, ecstasy, happiness, exuberance, and so forth, you could roughly rank them in terms of their adjacency, thereby creating a granular emotional spectrum of sorts. The odds are that not everyone would rank them quite the same, though there should be a reasonable coarse level of agreement. Even though a subjective judgment, it is not as if anyone is seriously going to put despair as an adjacent emotion between joy and happiness, for example. Therefore, we can see the possibility in principle, of approximately identifying and ranking definable emotions within a spectrum, based on their subjective qualitative evaluation.

If you could do that, you could then use them to code for meaning, just as different colours can code for a crude conceptual language, such as with the red, yellow, and green lights of traffic signals. Such an application is like the way that symbols and characters can code for much more complex languages. Some Chinese characters, for example, can code for several different descriptive phrases or images, the meanings of which are contextually – and in speech, tonally – selectable, and then strung together to form the compound phrase of the intended meaning. Another such code-for-meaning application may be the way numbers, symbols and letters can communicate through mathematical formulas and algorithms, which is also a language. The point is that it is not unreasonable to appreciate emotions as enabling communication if modulated in some unfamiliar way. Perhaps not *that* effectively from our point of view, though we cannot know that because we are not consciously privy to the process, just to some of the felt results. In the same way that the mammalian brain centre may broadly be considered to "think" and "communicate" in emotive terms, the later-model brain centre, the neocortex, has evolved to "think" and "communicate" mainly in terms of symbolic and encapsulated conceptual representations based on experience, and perhaps, in varying degrees, to both. Thus, we can appreciate the possibility of a communication method, a language of sorts, based on emotional values, which the nonlocal part of the individual unconscious might use for primary dialogue with a universally distributed consciousness.

I am not proposing that some grammatically structured communication takes place between an individual mind and a universal equivalent, in the sense of a transmission-stream resembling the emotional equivalent of symbolic characters. However, the intention here is to be open to the possibility that emotion is communicable, and in that communication, information may transfer, albeit in crude form. For example, that simple traffic signal application, with its three-colour-frequency coding, was arguably the forerunner of multiplexed light transmission through fibre-optic cables, which now provide communication of almost limitless information.

Emotional Maturation

If emotion were a language of spirit, how fluently might we dialogue? The answer is probably "inefficiently." In our defence, our species has become somewhat sidetracked by the alternative development of overt symbolic language. We are as a child coming to grips with newfound skills. We play and learn, and cast aside, then return to learn some more. As a society, we have been doing this spiritually during the "toddler" years of our evolution – the last few millennia. In psychology, the child-into-adult psychological "growth rings," though not sharply defined, pass through specific evolutionary stages toward maturity.

These stages of consciousness represent the individual or societal emotionally-based philosophical reference points or comfort centres for relating to life - their chief coping perspective. Briefly, borrowing from Ken Wilber's work on a philosophy of consciousness, these proposed stages of perspective can be ranked by increasing maturity level. These are:

- Archaic (might is right)
- Magic (hidden help)
- Mythic (superstitious ritual)
- Rational (logical reasoning)
- Pluralistic (non-egocentric care)
- Integral (non-discriminatory care)
- and eventually, Super-Integral (universal reverence)

In practice, it is quite clear that for every emotionally maturing individual and society, our mental and emotional evolutionary stages are quite patchy. Almost any subset of our society, such as a nation, tribe, family, or individual, will reflect the same patchiness of psychological, spiritual, and emotional development. We do not all develop in a stepwise manner, nor do we keep in step as we each develop. Some have even attempted to define a relative measure of this emotional maturity, terming it emotional quotient (EQ), based on the better-known concept of measuring intelligence – the intelligence quotient (IQ). Right across society's psycho-spiritual maturity distribution curve, you will find examples today of those still stuck in an Archaic coping mindset. There are plenty of examples of those remaining in the Magical and Mythical stages – perhaps rather too many for the health of our evolutionary "age" – though probably most are or would like to think they are at or around the Rational stage, most of the time.

Then, as the distribution curve heads down in numbers and becomes more elevated in psycho-spiritual maturity, we pass through the Rational portion to the soon-to-be spiritually awakened – most of whom are at the Pluralistic stage. Some examples of awakened leadership occur at the Integral stage, with a few, so very few, toward the leading edge of the curve, the Super-Integral stage. As society continues to evolve, becoming more aware, more familiar, more accepting of the language and perspective of spirit, the median of this distribution curve will gradually move forward. The bulk of humankind will gradually transcend Rational, then Pluralistic in turn, until, in some distant age, much of the species will centre around Integral and onward.

Much, if not all, of this psycho-spiritual development of emotional maturity, may depend on our awareness of, and nonconscious conversance with a universally expressed language of emotion, and our willingness to be open to its message. Here is the potential for awareness, and emotional centring in peace, love, and compassion, to amalgamate and form the necessary individual cornerstones for underpinning those same characteristics throughout human society. As far as meditation is concerned, there is no question that the emotional state of mind most beneficial to permitting a meditative experience is the white light of compassion, though in a gentle, benign, and somewhat detached form, not so much as burning passion. Reflecting the essence of "soul-consciousness," compassion is the emotional state of mind proclaimed by the great spiritual traditions as being the most sacred of emotional mental states – the mark of prophets, saints, and creators.

Ego

Ego turns out to be another of those words that we think we understand the meaning of, though on closer inspection, we do not. I will not attempt a clinical definition for such a complex subject, which is the terrain of psychology and psychiatry. Nevertheless, of relevance to meditators, the ego is a significant inhibitor, if not the *primary* inhibitor of access to undistorted spiritual insight. The word "ego" comes from Latin and is directly translatable as "I myself." It is a mental artifact of self-awareness and has no independent existence within the mind. Essentially an individual's mental construct of self, ego primarily defends and reinforces the individual's concept of self worth and separateness. An excessive ego focuses obsessively on revising its experiences of the past and projecting its existence into the future to protect its owner from ever appearing deficient. Despite apparent criticisms of it, in the context of meditation we should not consider ego as a foe to be vanquished, nor as wholly negative since it appears to be an evolutionary offshoot.

The somewhat outdated Freudian psychoanalytic model of the mind designates the id as representing full consciousness and things repressed by consciousness. The ego, seen as mostly conscious, manages our response to external realities. The super-ego, considered as only partially conscious, acts as the conscience or morality-judging internal observer, thus fulfilling some similar aspects attributed to the nonlocal element of the unconscious discussed in this book. A more straightforward definition of ego, borrowed from Eastern traditions, notably Buddhism, proposes that the ego maintains the mental illusion of independent and separate existence of self and others.

In the public eye, primarily influenced by today's popular self-help articles, ego may have developed a villainous identity of its own, like a melodramatic ham actor. Indeed, in excess, it can be the source of much suffering. It is not all negative news, though. Ego is, after all, primarily the recognition of self, and that, per se, is not a bad thing. Some moderate aspects of ego can help bolster an individual's assertiveness and self-esteem, placing them more confidently on the map of life, tiding them over the rough patches of external criticism or self-doubt. The chief concern here is the chronic, excessive form of hubris – unbridled ego of inflated pride or exaggerated projection of self worth often intended to disguise feared inadequacy. We are not trying to ban ego – as if that were ever possible – but we can bring to light and mitigate its potentially adverse influences on the mind's ability to meditate and to authentically interpret the resulting experience.

A well-balanced ego could restrict itself to alerting the mind to a real threat, then standing down if that threat were neutralized. In this role, it could just focus on one's self-awareness and support the development of a healthy sense of self worth, though the ego does not seem to limit itself this way except in some well-balanced individuals. For many of us, it prefers to remain on guard duty with a hair-trigger, defensively firing off at any perceived threat, real or imagined. The ego uses many tactics to maintain or improve the imagined status quo of the mind's desired self-image. Many of these involve ways of seeking, preserving, provoking, or enhancing validation from the outside world, as well as a host of less mature but recognizable tactics such as shedding personal responsibility, denial, redirection of blame, tantrums, invention, exhibitionism, sarcasm, lies, bluster, rage, drama, and even violence. These self-image defensive tactics may be triggered unconsciously, although ego selects those tactics based on the perceived nature of the threat, the maturity of the individual, and strongly favours repeating whatever worked before.

Characteristics of Ego

Understanding, recognizing, and overcoming the illusions and excesses of ego is a useful prerequisite to entering profound meditative states and authentically interpreting the contents of those experiences. All of us can think of extreme examples of individuals whose ego-driven response is so evident that it is virtually a caricature of who they would like to be. At one extreme are the aggressive, egotistical types, such as the hyperbolic salesperson, evangelist, or politician; these appear to be generally offensive strategies. At the other extreme are the defensive, retiring types – the acquiescent, the chronic victim or martyr. Common to both extremes is the capacity of denial and self-deception, the desire for some form of external validation or recognition, and the deflection of responsibility or blame away from oneself. Few of us, without some honest, inward contemplation, would be able to identify our full range of personal ego-driven responses, and to varying degrees, maybe we have all of them available to us as part of the human condition.

Frequently, ego may attempt to compensate for a sense of inadequacy, low self-esteem, or the absence of sufficiently robust self-validation, by seeking external validation. That is a coping response whereby more importance is attached to other people's opinions than to one's own self-assessment. Such individuals may identify exclusively with the reactions of others to them; they need convincing by others that their worth is more than that which they have privately assessed themselves. Any meditative endeavours under these circumstances may be influenced by fear of inadequacy, the desire to report interesting or entertaining experiences, and to be seen as doing well. There is a common thread running through many such symptoms – coping with low self-esteem – and this seems to be a fundamental aspect of the majority human condition. If ever humans have embedded within themselves an "original sin," as some would have you believe, it would not be inherited from Adam and Eve falling from God's grace; it would be a chronic fear of inadequacy.

Ego can play other roles. Chief among them might be storyteller and defender. The storyteller generates a constant background mental chatter relating revised versions of current and past events as they pertain to self while casting self in a more favourable light. Mental recordings, such as those learned from parents, teachers, or peers, may play back into the conscious mind about how clever or incapable we are or how we must avoid being wrong. If such influences appear to devalue self worth, the ego may provide a revisionist explanation denying responsibility. Our ego's chatter contributes most of the meditation-interfering "noise" that goes on in our conscious mind. If we could hear someone's ego chattering away, it might be amusing – until its blatant self-serving and facile nature became intolerable.

> *"Why are you unhappy?*
> *Because 99.9% of everything you think,*
> *And of everything you do,*
> *Is for yourself –*
> *And there isn't one."*
>
> Wei Wu Wei (Terence James Stannus Gray)

By exaggerating the role of moderate defender and stabilizing mechanism, the ego can develop as a defensive fortress, dedicated to denying entry to all potential threats, real and imagined. It can invest much effort into deflecting blame and responsibility, and denying perceived weakness, fragility, worthlessness, deficiency, or inadequacy. Ego may eventually convince some that they can no longer afford to expose their true self to anyone, including themselves, and thus must continually maintain a pretense. Therefore, they must have a story – a persona – and ego will gladly provide that persona. It will practise it, embellish it, and repeat it by heart *ad nauseam*, and once more, the real self becomes sacrificed, interred deeper inside that continually reinforced fortress. The problem with a mighty fort is that while it may keep perceived threats out, it also traps the defended within, eventually becoming indistinguishable from a prison. Thus, the protected owner of ego – the true self – finds itself becoming the incarcerated tenant, never daring to experience authentic, unconditioned freedom. If these engorged strategies appear successful, they become used to offset lesser threats, until an overactive ego is on full-time guard duty, effectively becoming a permanent facet of the troubled individual's character – and authentic self disappears.

Ego Quieting

If we are open to it, meditation can help us to venture outside an ego-induced fortress, to rediscover our real self, and to know our existence and reason for being. When first starting self-perusing meditation, one might feel some apprehension over any naked revelations of self-discovery. Possibly, childhood feelings of inadequacy might resurface. The desire to avoid these feelings might trigger the ego to come barrelling out in an aggressive exhibition of denial and defence, creating tension, and effectively stifling authentic meditation. The key lies with openness, gentle meditative practice, and the choice at the outset of a low-profile attitude towards spiritual exploration – the beginner mind – as mediated by authentic intent, humility, and mindfulness. In this context authentic means the degree to which one is true to one's original being and spirit, despite external influences. To meditate effectively, we may need to heal that which requires an aggressive ego.

Ego need not fear outright banishment, however. It can be brought gently into conscious awareness through iterations of contemplation and meditation. There, we can recognize it for what and why it is; self-esteem issues can be identified and gradually healed, and the ego's neediness diminished. The intent of authenticity in meditation is to be genuinely open to self-directed, non-narcissistic loving-kindness and new spiritual discoveries – to be genuine, period – and to have the humility to approach the acquiring of insights with innocence unconditioned by prejudice. We can be gentle and forgiving with our thought processes and with ourselves. Quieting ego's noise is not a battle – that would be like the left hand fighting the right – it is more a subtle interplay of awareness and intent. Ideally, we need to be in that mental state of soul-consciousness, when we can quiet the mind, sit in tranquil meditation, find our authentic voice based on the awareness of our small part in a unified universe, and know who we are.

Perception

One might think that when we examine our reality, what we perceive is what is there. However, it is illusion. As Lao-Tzu commented in his Tao Te Ching, twenty-six hundred years ago, *"What is comes from what is not."* I am sure he was not referring to quantum mechanics, nevertheless, his philosophy appears spot-on at the subatomic levels of our reality and can scale for all levels of perception. Since perception influences thought, it can affect the quality of meditation.

Spiritual awakening encompasses the ability to commune with one's inner self while coping with the outer "real" world and recognizing that one's authentic internal perceptions, though subjective, are not so much different from the outer ones. Granted, physical interaction in the real world, such as running into a wall, can seem a good deal more tangible than most inner perceptions. However, that apparent dualism between inner and outer realities is questionable, and we may discover that they are not mutually exclusive but complementary, part of a shared continuum of existence separated only by how we interact with them.

Physics of Perception

Once thought of as being solid, indivisible, and the smallest, most fundamental physical entity, we now know that atoms consist mostly of space, with some concentric clouds of electrons orbiting around a tiny nucleus core made up of protons and maybe some neutrons. Protons and neutrons themselves are also mostly space, containing triplet combinations of quarks held together by nuclear forces, and vibrating incessantly.

Moreover, those fundamental subatomic particles, such as electrons and quarks, are not particles at all. They may not even be fundamental, and the term "particle" is just a convention. They are all just various fluctuations in universally distributed energy fields. They are quanta of energy in indeterminable, probabilistically influenced states of potential, masquerading explicitly as both energy waves and as particles of matter. Even the strong nuclear forces binding those "particles" together are just exchanges of energy. Thus, subatomic matter is little more than space – over 99.9999 percent unoccupied – interspersed with a smattering of turbulent energy vortices; no solid objects at all. Conceptually, you can consider matter as an extremely condensed form of localized vibrating energy fields. Furthermore, at the subatomic level, the state of that condensed energy is inherently probabilistic – its exact state cannot be determined. Matter is virtually nothing except arising probabilistic waves of energetic turbulence, much as Lao-Tzu had supposed.

That represents a challenge for the perception of solidity in our outer reality. How do we, who consist of the same material, navigate through this swirling field of quasi-present, probabilistic, vibrating energies we call reality? We can only view its appearance, feel its solidness, hear its sounds, smell its fragrance, or taste it through our senses or some extension of them, because we, our senses and anything else we could use to probe it, all comprise the same vacuous substance. Furthermore, one can also question the reliability of those senses.

Consider, for example, the reliability of our visual sense. Imagine you were in a field of bright red flower blossoms. The surface cells of each petal contain an abundance of atoms that this flower's genome specified to create the colour red. The red-making atoms in those surface cells effectively absorb and convert energy from the full light spectrum received and then emit a portion of it as red photons. Thus, the red flower can emit more red photons than it receives – hence its vividness – and some of those photons head toward your eye.

A photon enters the eye, making its way to the retina at the back. The retina contains over 100 million sensors, capable of detecting light, colour, and relative intensity. These sensors convert the photons into electrochemical impulses, which travel to the brain through the optic nerve. The brain feeds this intensity and colour information through its visual processing centre to part of the unconscious mind, which processes and stores the information, making what sense of it that it can. If, and only if, the unconscious considers this information worth sending to the conscious levels of the mind would it do that, around a half-second later.

Supposing the image received unconsciously were not just of a flower but included a charging bull. The unconscious brain would immediately activate various glands and motor muscles to galvanize the body into flight mode, a full half-second before the conscious mind even becomes informed – not of the flowers, but of the bull. Thus, by the time one became consciously aware that the bull was charging, one's body would already be more than a half-second ahead of any conscious decision to run.

One might think that such a delay in reaction would be noticeable. However, our consciousness gets and gives its body status and command messages through the unconscious, subject to this same built-in delay, so it does not realize the body is already starting to run before it has consciously decided to do that. With such built-in delays, it is surprising we can consciously relate to our reality at all. A mitigating factor, which enables us to navigate this world from a perspective continually a half-second out-of-date, is our hard-wired predictive ability.

Unconscious processing continuously expends much effort in extrapolating and anticipating based on the wealth of sensory data at its disposal. This predictive ability enables us to anticipate where a thrown ball will be by the time we can offer up our hand to catch it. It allows us to navigate our way down a flight of stairs or modify our running stride to leap the gate before the mad bull arrives. Overall, this unconscious predictive ability is surprisingly accurate in the short term and is a vital part of our survival heritage. The leading operational presence in charge of our body, particularly in pressing survival matters, is the unconscious, with consciousness having a minor supporting role – mostly of playing catch-up. Consciousness can generally only process one thing at a time, whereas the unconscious can process data from all the senses. Research shows that the unconscious sensory data-processing capacity is around 160,000 times greater than that of the conscious.

Returning to the red flower, the unconscious has registered the red petals and may or may not inform consciousness of it. If the red petal photons happened to enter your eye while you were scanning the field trying to locate and avoid the mad bull, your unconscious would be unlikely to inform the conscious you of the beautiful flower colouring. If, however, there were no mad bull and you were consciously contemplating the beauties of nature or looking for a photographic subject, your unconscious would most likely oblige by bringing the red petals to your conscious attention. The wonder of it is that unless something threatens the self, or there is deliberate intent – a mindful, focused awareness – the vast majority of what our eyes see never even enters the conscious mind.

Much the same is true for the other senses. If you stub your toe on a rock, a chain of energy exchanges takes place between the atoms of the rock face, your boot, your toes, then through your nervous system and brain for unconscious processing. Unless it is severe trauma, or your safety is compromised in some way, your unconscious mind may not even inform your consciousness of the event – particularly, perhaps, if you were simultaneously fully engaged in a higher priority activity, such as evading a charging bull. One might conclude that consciously we go through each day with blinders on while watching low-resolution replays of past events, and that seems part of our design. It has been likened to driving a car forward by only looking in the rear-view mirror.

Perspectival Influences

Humans have been viewing their surroundings for hundreds of thousands of years. Still, the appreciation of perspective is a relatively modern development, one just a few centuries old. However, perspective does not only apply to artists or the perceived geometrical relationship of material things. Differences in viewpoints of a philosophical, attitudinal, cultural, or psychological perspective can also influence and distort perceptions, understandings, and concepts.

As it might affect meditation, perspective is how one regards situations or topics; these may be skewed by mental filters installed due to first-hand experiences or those reported by others. At a societal level, such prejudices can produce restrictive cultural paradigms such as archaic customs extending into ritual and superstition. Therefore, it is likely that historical descriptions of ancient events could be coloured by the then current standards and conceptual perspectives – the cultural status quo – and subsequently by new paradigms revising the retelling of them. Future translations of those ancient descriptions into new languages of the day, intended to clarify their substance corresponding with contemporary perspectives, would also provide the opportunity to insert cultural, political, and ego-driven editorials, distortions, or fabrications that were not present in the original narrative.

Consider, for example, the dramatic phrase "a fiery chariot of the gods." If recorded a few thousand years ago, that might well have been the best available portrayal of a natural event such as a meteor, by a person living in that time. They had neither a word for nor even the concept of a meteor. The fastest large object they were familiar with might have been a chariot. The meteor's burst of light and trail of smoke, and accompanying flash and explosion if it hit the ground, would undoubtedly suggest fire. Moreover, the most probable current cultural paradigm of the day would be that the sky and the heavens were one, and the domain of the gods. In these circumstances, the entire description is rational by the levels of conception and cultural perspectives of that time.

Furthermore, one can imagine that during the retelling or translation of this relatively primitive description of a natural event, into whatever was the next favoured modern language and paradigm, the opportunity existed to embellish it. Overly motivated translators and scribes, mostly religiously educated monks and clerics, could creatively equate the event to "a sign from God," and even add in a cast of supporting characters. A few more centuries of self-serving editorial translation and revisionist retelling could portray that single event as an underlying narrative supporting the foundation of their religion. Therefore, we may acknowledge the plausibility of a distorting effect that a society's contemporary, paradigm-influenced perspectives can have on the recording and retelling of ancient events and learnings.

In trying to identify the underlying bases for the spectrum of recorded spiritual events and revelations allegedly experienced by humanity and subsequently inscribed into the historical record, we might conclude that there is no clear way to anticipate and compensate for such distortions. Some might thus choose to acquiesce to the dictates of a third party claiming the only way to find truth is by following their understandings by rote and communing with their god through their structured belief system and rituals. However, if we do that, we are effectively giving up our spiritual power and self-responsibility to that third party and their politics.

In the willingness to cede one's spiritual control and shed responsibility lies the basis of superstition, dogma, ritual, and worship. It can lead to the unquestioning following of prophets and princes, seers and sovereigns, and the surrender of rationality to belief in myth, magic, and miracle. Individually or societally, this may manifest in the invention of mythical beings, onto whom one can project one's hopes, fears, and fantasies, and to whom one can ultimately surrender all one's responsibilities – the pre-rational mentality. Therefore, if one expects to see dragons or divinities during meditation, then perhaps that is all one will see.

Ultimately, though, perhaps the only solution for an authentic spiritual understanding is not to blindly conform to the dogma of others, but to investigate for oneself. If we choose to conduct such an investigation through internal enquiry by way of meditation and contemplation, we should be aware that in trying to avoid distortions from external sources, we might inadvertently create our own internal aberrations of perspective, which could influence any personal spiritual insights. I am not referring to conscious embellishment or outright fabrication, but to the subtle effects that one's unconscious biases and intentions can have on how the witnessing and processing of an event, be it external or internal, presents to the conscious mind.

A startling example of the power of unconscious psychological influences on perception is a deceptively simple cognitive science study conducted several decades ago at the University of Illinois. Termed "Sustained inattentional blindness," or more colloquially, selective attention, the study comprised two groups of three people, one group dressed in dark clothing, the other group in white, continually passing a basketball among themselves on a random basis. Observers were tasked to count the number of passes made by those dressed in white. Partway through the study, a person dressed in a dark gorilla suit walked through the group, paused, pounded his chest, and then walked off. He was visible for nine seconds. Less than half of the observers noticed the gorilla. The study was repeated with mildly inebriated observers, and less than one in five acknowledged it.

These results dramatically point to the unconscious mind's ability to suppress everything except those events deemed important enough to report to the conscious mind – in this case, what the white-dressed people were doing. This effect is an actual example of the same response we discussed previously, in which the focus of unconscious awareness on a charging bull out-prioritizes and precludes the conscious from becoming aware of red flowers and stubbed toes. The unconscious mind can play tricks on one's consciousness in many other ways; tricks potentially even more acute than external reality events like the "gorilla in their midst" experiment just described.

Our unconscious responses to the world can influence the quality of our meditative experiences. For example, phobias or prejudices can become habituated into the way we think, creating a permanent warping lens introduced between the world and our perception of it. Perception typically means becoming aware of something through one or more of the senses and is a fundamental component in the understanding and formation of a concept. In contemplating the meditative experience, we are more interested in the internal representation of what the mind is witnessing – awareness arising through the sense of thought. Prior experience and imposed conditioning can develop preconceptions, and these may influence the processing of novel or traumatic mental experiences.

If, for example, one has a fear of snakes, then in a darkened room, a coil of rope might appear to be a snake. However, this example is of a physical object directly arousing our senses and being momentarily misconstrued by the prejudices of the mind. Material objects generally remain around long enough for a physical second look, permitting other senses and the rational mind to determine what they really might be. Perceptions originating in the mind, however, have a greater potential for unfalsifiable misrepresentation. When perceiving something as subjective as ephemeral flights of conscious insight, their inherent transience denies that second look. Thus, it is prudent at the outset of meditation:

- To relax and clear one's mind of sources of perceptive distortion such as tension, prejudices, and coping mechanisms

- To be prepared to be open, without judgment

- In all humility, to proceed to experience the condition of being

We do not have an adequate vocabulary to describe inner perceptions. People may talk of observing thoughts, images, visions, dreams, or even hearing voices, but these are words borrowed from physical perceptions of the tactile external world. How, for example, does an idea present itself – it is not really a lightbulb, is it? I imagine that reception or formulation occurs nonconsciously, which then unfurls, rendering itself into conscious awareness as a realization. During contemplation, one can sometimes feel this formulation of an insight happening in slow motion, as it occurs within the mind. The process seems to follow a similar route to that of sensory awareness, where the unconscious mulls the information received from one or more of the physical senses for a few hundred milliseconds before deciding, based on its determination of salience, whether to push the information onward toward consciousness or not.

With an idea, the information is garnered through nonconscious processes from local or nonlocal sources, instead of the physical senses. As in dreams, the thought patterns pushed by the unconscious toward the conscious mind are disseminated throughout the brain's various applicable processing faculties; the same ones that would deal with an actual physical stimulation. These include short-term memory, conceptualization, and perhaps one or more of the sensory information-processing faculties. Among those faculties relevant to the thought-content in question, a synchronous resonance develops over a few hundred milliseconds, effectively reifying the unconscious thought into conscious awareness. As with sensory information, a half-second or so can elapse between an idea's arousal in nonconscious processes, and the conscious realization of the occurrence – the blooming of an insight.

A calm, conscious mind, having perceived the intrusion from nonconscious processes, creates a representation of it to aid in memorizing and later processing by selected faculties of the brain. The nature of this abstract representation and its subsequent conceptualization can be influenced by expectations, previous experiences, one's inventory of relevant knowledge and analogy, and by cultural or learned prejudices. This condition underscores the challenge of effectively representing an event for which the mind may have no prior experience, no adequate conceptual models on which to lean, and no existing vocabulary to describe – when the meteor can only be described as a fiery chariot of the gods.

Meditative practices mostly develop quite delicately, though sometimes they will suddenly allow us a glimpse of the underlying nature of existence. Such a flash of insight may be incomprehensible to us due to our limited abilities but can still provide directional or even profound clarity into existential matters. Such spiritual lucidity or awareness is often accompanied by a feeling of intense confidence in its truth. This subjective state has been recognized over the millennia as untaught or intuitive knowing rather than belief-in-narrative, and the ancients referred to it as *jnana* in Eastern culture or *gnosis* in Western culture. *Jnana,* a Sanskrit word used in both Buddhism and Hinduism, means conscious knowledge or realization typically acquired through meditation. *Gnosis*, an ancient Greek word, was used by European-based religions to describe intuitive cognition of a divine nature. They both describe a cognitive event that, when experienced, is recognized as intuited knowledge inseparable from the experience of reality.

Let us now take a closer look at what is generally regarded as reality.

INSCRUTABLE REALITY

The journey described in these pages is about trying to look beyond accepted paradigms, beyond the illusion of our "reality." The challenge is to look behind the Wizard's curtain, behind the theatrical scenery, to see the world of phenomena for what it is, and perhaps to discover deeper truths about the universe, existence, and ourselves. We may embrace the opportunity to understand in our hearts what our minds cannot fathom – how we relate to each other and all that there is. With such insight comes the knowledge to live our highest good, all the better to fulfill our destiny. Here, we will briefly reflect on some of those intangible though physical elements of our existence and question common assumptions about their nature. An appreciation of this background will put us in a better position to follow later discussion on the nature of existence.

Time

When pressed for an understanding of the nature of time some fifteen centuries ago, Saint Augustine of Hippo made this observation.

> *"What then is time? Provided no one asks me, i know.*
> *If i want to explain it to an inquirer, i do not know."*

Much later, Einstein commented laconically that *"time is something you read off a clock."* Our obsession with time was enabled by the invention of the clock in its many forms – from sundials to atomic chronometers. Before clocks, we had a much more leisurely temporal apportionment system comprising seasons, weeks, and generous portions of days and nights related mainly to the disposition of the sun, the anticipation of the dawn, or the growling of the stomach.

We tend to talk glibly about time with a familiarity portraying comprehension. It seems a critical factor in how we exist. Time can be a mysterious, romantic notion, and it is possibly the most deceptive "dimension" of all. It may not even be a dimension. Many metaphorical descriptions of time exist, which compound the mystery. "Time's arrow" implies that time follows a one-way sequence from the past, through the present and into the future, and cannot be reversed. A poetic metaphor of the "flow of time" is like a river, where our metaphoric perspective would be from the riverbank. We look upstream to the future with anticipation, sometimes fear; downstream to the past with satisfaction, sometimes regret; and across mid-stream in confusion, wondering if perhaps the other side might provide a better vantage point.

Another classical time analogy is that of older-technology film-based movies. Anyone who has seen a film-reel mounted on an old-style projector will know that it is a continuous strip of cellulose, on which a sequential series of still, semitransparent pictures appear. As the film moves through the projector mechanism, a bright light projects the image of each still frame in turn onto the screen. This sequence of still images shows the change in state from each replaced picture, which, when viewed in quick succession, is consciously interpreted as a smoothly flowing animated image, changing with respect to the passing of "reel" time.

Similarly, a model for each moment of the present might be the experience of a change in the quantum state of every particle in the universe from the previous state. If these changes were fast enough, minute enough, and in a correlative sequence, they would create, in any observer's mental awareness, the illusion of a smoothly flowing, time-based continuum unfolding. We would feel that this sequence of differences evidences the flow of reality along time's axis. The concepts of time as a temporal dimension, like a continuously flowing river, or as an arrow, is helpful in our ability to model and manage our daily physical existence. However, just as with any model, these are ultimately flawed.

Some mathematical formulae used in physics, particularly quantum physics, imply that time's arrow is reversible – at least when reduced to the subatomic level. However, the most probable direction of change in time at the macro-scale existence is toward disorder, converting to a lower energy form. This trend is termed entropy. Even at the quantum level of particles, the most likely change of highest probability – generally associated with least energy expenditure – will evolve toward a lower energy and more disordered form, thus increasing entropy. The net effect of this is that multiple interacting particles will tend to change in that same temporal direction because it requires less energy and is statistically more probable. After all, yolks rarely rise off the floor and reassemble themselves into an egg rolling back onto the countertop. Reversal of time would imply a process of regaining unimposed energy or order, thereby undoing entropic events. The probability of such a spontaneous process occurring at random may not be entirely zero, but it is unimaginably low. The one-way nature of the arrow of time results from a statistically cumulative effect of a spacetime network of microscopic, probability-driven state-changes.

In practical terms, a macro-scale dimension of time into the past is not physically available to us. There can be an imposed reversal of process to recreate what was, but that process takes place in the present, and with some net energy expenditure. It would not lead to the past, just toward a different future, one that emulates expired versions of the events involved. At the macro scale, no intrinsic past exists for access by us. Nevertheless, such a concept is a useful one, and used in math models – and science fiction – extensively by us sentient beings. It has worked its way into our thoughts and language so much that it is difficult to conceive otherwise. Knowledge of the past acts like a map – a reproduction of what once was, a historical trail of used-up present states – which helps us project into and navigate this world; but the map is not the terrain. The terrain is the Now, an instant emerging from a foundation of spent residues of all previous versions.

Neither is there an accessible or determinable future. A potential change from our present state can be anticipated, and could even be highly likely, but it may not happen. It ultimately emerges from cumulative probabilistic events at the quantum level. Heavily weighted statistical probabilities can underpin the prediction that an event is most likely to occur, and which we might even deem as inevitable. Nevertheless, it is not – even a new day dawning has a small uncertainty associated with it. Future possibilities might be put in place by macro events, including human intervention, or by other influences at the quantum level. However, their occurrence is never certain until experienced, and by then, it is already the receding past.

No pre-determined, pre-ordained future or fate exists; nothing that is already set up awaiting our temporal arrival. There is no forward-looking certainty. The universe and its sentient occupants are free to invent or circumvent a malleable future by the physical application of energy, or any other means of influencing the probabilities of entangled, superpositioned, potential quantum events. Nothing is sure to change its state specifically in our next version of the present, yet everything will. The future state of such change will be probability-driven, as all past states were. We can huff and we can puff, but all we can ever hope to effect is a tiny slice of our local existence in the present, by putting in place changes that may or may not cause other changes that may or may not lead to the desired outcome. Such an intended future state may only result provided the balance of all its linkages to the rest of the universe — all other "uncontrolled" ongoing changes and collateral events — are insignificant or supportive influences.

In a looser sense, however, we can consider time-like hops into a future, through the effects of relativity. Merely by occupying a spaceship capable of accelerating to a significant fraction of lightspeed and flying for hours or days of on-board time at maximum power, you could return to this location to find days, decades, or even millennia have passed locally. The relative duration of the local time that has passed would be dependent solely on the lightspeed-approaching ability of your rocket, and for how long, in terms of ship time, you stayed on that relativistic journey. In effect, the faster you go, the slower your time appears to pass for someone left behind at your departure point. Conversely, from your perspective in the rocket ship, time at your original departure point passes faster.

This form of relativistic "time-travel" scenario does exist in practice. For example, signals from the earth-orbiting Global Positioning System (GPS) satellites that we use for navigation, need continuous minor adjustments to account for small relativistic effects due to the satellite's motion around the earth. However, like life and hibernation, this relativistic mode of time travel to the future is strictly one-way. You cannot undo it; there is no going back. Having leapfrogged into a far future in this way, all your past relationships, your old life, including everything occurring since your departure has passed, dissolved. You are irrevocably committed to your new present – in our future. Enjoy!

Supposing that time is not a dimension at all, where it has no intrinsic existence in any inherent form. Ancient Eastern sages refer to the "four moments," these being:

- The immediate past

- The conscious present – what we might call the Now

- The imminent future

- A timeless state revealed in deep meditation – an awareness of a limitless expanded "Now" with no temporal boundaries

In this context, the cliché "there is no time like the present" would be a truism. What we sense as the passing of time is just the awareness of a cumulative history of multiple sequential occurrences of changed Now states. The perception of time elapsing is then the interval between such changes, and the duration of that interval appears limited by the width of what must be an extremely narrow "window," a momentary sampling of the experience of a present reality. The width of that window would be the temporal distance between the trailing edge of Now – the immediate "past," and the leading edge of Now – the probabilistic imminent "future." When that future does become Now, it is no longer probabilistic. Its many multi-possibility states have collapsed to become the present event.

What we call the past is but the echo of completed changes, stacked in a succession of expired Nows. The material debris of those echoes may retain some form of reality, as can be evidenced through sentient memories, or surviving physical records of historical changes that have occurred – whether over microseconds, millennia, or eons ago – and which may survive into this Now. Nevertheless, these records are limited to the residue of changes already completed in previously expended Nows. Although they may exist in this Now, nothing exists in the past, because there is none. Nevertheless, these granular crumbs of evidence do provide a solid basis that instinct or the inventive mind can use to map a projected path, an extended sequence anchored in the sure history of prior Nows and wavering toward an uncertain future.

This description of time as a history of changes leading up to the present can be viewed as an always-on transition between a just-dissolving state and a newly manifesting one. Now is but an instant, a flash coinciding with some manifestation interaction. Ironically, the instant an uncertain future has manifested into an experienced present, it is already a dissolved past version. However, our mind appears to have a slowing-down mechanism in the form of recursive refresh loops, which effectively thicken the apparent duration of the perceived Now. Thus, our awareness can better react to transitory events, even though they have already been completed, and can use that data to project ourselves into an uncertain near-term future. In the manifest physical world, the only time we experience is through this narrow window of dilated Now – an instant of unconscious present. Consciously, we are permanently living in a state of awareness of previous Nows – in the past. We survive by projecting our intended actions some half-second into the future. It does not always work! There is no conscious window on the present, of any duration, for we already predict and operate based on passive information provided by expired Nows.

By making use of the natural relationships between known physical constants, including lightspeed, the physicist Max Planck calculated the minimum "temporal length" for the spacetime granularity of the universe. He designated this minimum dimension as Planck time. Essentially, this represents the finest resolution of time – the realizable limits – below which nothing could interact with our reality.

Planck time may, therefore, represent the temporal width of Now – an extremely short interval of some 10^{-43} seconds. (We will use the mathematical notation herein as shorthand to describe exceptionally large or exceedingly small numbers. For example, 10^{12} means 10 followed by 12 zeros – or 1 trillion; and 10^{-12} means 1 divided by 10 followed by 12 zeros – or 1 trillionth. So, 10^{-43} seconds is the math notation for a ten-million, trillion, trillion, trillionths of a second.) Although this is a tiny number, the point here is not its numerical value. It is the principle that there might even *be* such a finite discrete temporal resolution – a quantized granularity of time made up of accumulated "Now" windows – responsible for the apparently smooth temporal continuum of our reality.

Nothing can travel faster than light. Not only light, but the entire electromagnetic spectrum has the same finite speed limit throughout spacetime that cannot be exceeded. Gravitational force effects, which are quite different from light in that they relate to mass-energy distortions of spacetime, also have the same transmission speed limitations. These common limitations led to the conjecture – confirmed recently – that this speed limit is not a property of light or gravity itself. It results from the universe being unable to react faster to *any* interaction or physical change – it is a dynamic systemic response limitation of the universe system. It has a granular limitation in its ability to restructure itself in response to any experienced stimulation that would result in a change to a new Now. Thus, one must conclude that the granular concept of time results from the universe's limited ability to accommodate structural change.

Change

Based on our macro experience, we might imagine a local subatomic change as a process that transitions between beginning and end states in a smooth continuum over some small amount of time. However, at the quantum level, change is quantized. There is only state A or state B, for example. No in-between, no mid-state, no gradient, or any other type of transitional existence intervenes. Like the consecutive frames in that film-based movie analogy mentioned earlier, you either have frame A, or you have frame B, and it makes no sense to ask what the actors do in-between frames. Such film-based movies are effectively quantized at the level of each frame. There is no in-between frame transitioning existence. A state that was and then is not, is replaced by a new state that was not, and now is. That represents the absolute granular resolution of the process. At the microscopic level, reality switches from one state to another, with no in-between existence, the change being an instantaneous shift with the new reality instantly overwriting the previous one.

The perception of time passing seems linked to the local observer subjectively experiencing – and perhaps mentally dilating – the interval between sequential changes. One might imagine every quantum state in the universe being sequentially "repainted" for every new version of the present Now. That would be like a giant television screen, where every pixel refreshes many times a second to keep the images smoothly flowing. However, a moment's reflection will suggest that perhaps this analogy is misleading. It assumes a universality of effect – the entire universe changing uniformly. However, acknowledging all the distortions and variations possible in the spatial and temporal texture of the universe, we realize that there can be no uniformity to change – no absolute time-keeping pulse synchronizing every change in the universe, like some gigantic army marching in step. Any particle's status refresh rate cannot be a universal absolute since it only occurs with respect to the local frame of Now in which it manifests.

Thus, the notion of "Now" turns out to be not so much a single snapshot, or a clean cross-section of the entire spacetime universe, but a mesh of local perceptions. The perception of Now is relative and subjective, drawn independently by each sentient consciousness from spacetime, though forming part of a locally shared consensus. It is individual conscious awareness. The Now we become consciously aware of is our local perspective on a cumulative summation of the states of all prior local events – the history of changes up to the present as it was some half-second ago. Collectively, it is the endpoint of all cumulated histories, as observed from every subjective viewpoint of sentience.

We have noted that no real past exists, only the history of finished changed states. We also saw that there is no pre-existent future, waiting for us to enter. Now, we find there is not even a common present – no objective, absolute, and universally experienceable Now. Depending on the state of our local frame of reference relative to others, the accumulation of all the probability-influenced random quantum level activities provides us with the subjective, consciousness-related, and localized sense of Now – the illusory present of our macro reality. The apparent substantialness of Now only seems real if there are endless quantum level changes and the sum of their histories is retained; and there is no close examination of the process. Although we make good use of the concept of flowing time in our daily lives, change is the more fundamental characteristic.

Energy

Conventional energy comes to us in a handful of categories that distinguishes them and gives the impression of comprehension. Some traditional descriptions of energy include:

- Kinetic
- Potential

- Electromagnetic
- Thermal
- Chemical
- Nuclear

And more recently, these would also include:

- Rest-mass (of $E=mc^2$ fame)
- Vacuum energy

Within a closed system, the principle of conservation of energy is one of the fundamental laws of thermodynamics, stipulating that energy can be neither created nor destroyed. However, it can move around and transform from one form to another. During the era of the industrial revolution, human society became quite adept with inventions such as the heat engine (steam, diesel, gas, etc.); at getting work done by locally transforming energy from one form into a less useful one; and by exporting or importing energy components to and from the surrounding environment.

Some energy differential must always exist so that energy can flow from one state to a lesser one in order to do work. The least useful and final form of energy is ambient heat, which cannot create work by flowing toward a lower form, for there is none. The term for this tendency of energy to move toward its lowest ambient form is entropy. It is to this basement of the energy hierarchy that all energy forms eventually migrate. However, many sources of energy differential remain, which natural physics, random quantum events, complexity, life, and intelligent sentient management can tap into to produce work, create ordered states, and reduce or reverse local entropy – at least temporarily.

Despite what we claim to know about the various forms of energy, and how to make use of most of them, in truth, we do not know what energy is. Conventional classical science definitions, such as the "capacity to do work," "impetus behind an activity," or "ability to exert power," do not comprehend what energy is, just what it can do. That is defining something by its function, not its form, like describing a pencil as a writing stick – not that helpful. However, perhaps this is the only way we can conceive of energy – as just an agitative quality associated with the process of change. When expended, energy performs work by increasing local entropy, shifting from one level to a lower one. When generated through imposed change, it is the reverse. Energy and change could represent two sides of the same coin, with their sequenced causal relationship providing the experiential illusion we call time.

Dimensions

Three conventional spatial dimensions define our familiar physical world, along with time, which we noted as probably not a dimension. Descartes' concept of three physical dimensions, each orthogonal to the other – length, breadth, and height – to define the location of a point in space, led to the so-called Cartesian coordinates, which functionally remain fundamental to science and technology today. If asked, most people would say that the structure of our reality comprises three spatial dimensions, and maybe a fourth temporal one. Others, taking note of Einstein's concepts, might suggest just one – a volumetric dimension that evolves with time: the spacetime continuum. A theoretical physicist familiar with string theory – the latest candidate for a grand unification theory of everything – might say we exist in ten or eleven dimensions, and not so long ago it was thought to be twenty-six. However, that physicist might add, most of these additional dimensions are inaccessible and undetectable by us, so they must be hidden in some way.

Our conventional concept of dimensions is just that – a concept; it is another incomplete model. One we are so familiar with, that, confusing the map with the terrain, we tend to believe them to be actual reality. Conventionally, dimensions are degrees of freedom, directions in which something can extend, move, or be measured. They are not some invisible walls of a container within which we live. A so-called two-dimensional plane, while being a useful mathematical concept, cannot physically exist in our reality without the third spatial dimension, for even if that plane were only one atom thick, the very atoms forming its surface must occupy a volumetric space. Every subatomic element making up each atom requires a spatial volume in which to exist, move and vibrate. Spacetime provides the volume in which to be, formed by the three conventional spatial dimensions, and temporal intervals enabling change, motion, interaction, and evolution. Matter requires a volumetric space to occupy, particle by energetic particle, and time is the sequence of that agitation. We, therefore, live in the most fundamental dimension set that can support physical existence. The atomic structure of our physical reality can only exist in this single, minimum dimension we call spacetime.

When we talk of curved spacetime in discussions about the nature of Einsteinian gravity, one might wonder what the additional dimension would be, within which our spacetime curves. One might imagine an extradimensional hyperspace, within which our fundamental spacetime could flex, operate, and evolve. Conventional thinking assumes a fourth conventional spatial dimension to create this hyperspace – "the fifth dimension" – but a single conventional dimension might not suffice to accommodate a flexing volume. It may take an additional volumetric-like dimension, or higher, into which spacetime can curve. Whatever that extradimensional hyperspace looks like, our volumetric spacetime-dimensioned universe would exist within it, perhaps like a spacetime patch on the "surface" of a higher dimensional bubble.

The point here is to be open to the practical possibility of there being additional and different dimension sets involved in our existence than the conventional three-dimensional model. Einstein introduced us to the unified single "dimension" concept, the spacetime continuum evolving with time, which we can call dimension one. Then, any additional dimensions, to which we appear not to have access, might only exist in "quantized" units equivalent to our baseline volumetric space. That extra volumetric dimension, within which our spacetime flexes, but is somehow segregated from ours, would then represent a six-dimensional spatial model in conventional terms. The addition of a third volumetric extra-dimension, perhaps to accommodate the other two, would then equate to at least nine conventional dimensions, as a minimum. If you were to include the common element of time to enable change, a ten-conventionally-dimensioned hyperspace, as reflected in contemporary string theory, is readily imaginable.

Currently, there is no knowing in what form any such extra-dimensions might exist. Suffice it to say, some combination of additional dimensions is proving to be theoretically necessary to explain our reality. Though these extra-dimensions may seem to not interact with ours, or at least not be detectable by us in any conventional manner, they would likely contain or be contained by ours. However, there is at least one possible universal mechanism that we are aware of, which might bridge or leak between any such extra-dimensions, and which would lend support to this quantized hyperspace concept – the gravitational field.

Fields

The term "field" originally described an entity that spread across a surface, such as a farmer's field of wheat. Relating to physics and technology, it has evolved to model a set of variable values or characteristics at every point in a region of space. It can describe a networked effect, a matrix or flux spread throughout the spacetime dimension of which we are aware – and perhaps others of which we are not. We talk of electromagnetic fields, meaning a local volume of spacetime saturated with levels of electromagnetic energy, or of the short-range but intense nuclear force-field, which binds subatomic matter together. This usage represents a spatial volume, saturated with a flux modulated by the presence of particles, or causing them to be. Recent models of quantum field physics conceive of these particles as being extremely small-scale local oscillatory disturbances. They represent energetic nodes of vibration or fluctuation in an otherwise uniform energy field. Different frequencies and patterns of fluctuation may then produce the observed characteristics interpreted as being the various particles.

Some of these space-saturating fields may be bounded within our known spacetime dimension. Electro-magnetism and nuclear forces are probably so limited. Others, such as gravity and quantum fields, may saturate our spacetime dimension, but extend beyond it – not limited to or by our known dimension set. Our spacetime dimension may then figuratively float within such extradimensional fields. Other possible candidates for extradimensional fields include the vacuum energy, which is a quantum energy field, and the Higgs boson, a quantum excitation of the Higgs field, which confers mass onto all other particles, except for massless particles such as photons.

Gravity

Gravity is different from other forces. The conventional conceptual description is like a force field – a volume of space saturated by the forceful effect of mass-induced gravity. It can also be modelled as a flux of theoretical gravitational "particles" called gravitons. However, gravity is the only mono-polar force, all others being bipolar, like a bar magnet with north and south poles or an electric field with positive and negative polarities. With gravity, as far as we know in our spacetime dimension, there is only one pole – that of a mass attractor. Gravity, as Einstein contemplated it, results in and from the warping of spacetime through the localized presence of concentrated forms of mass-energy. Thus, a mass-generated gravitational field warps the structure of spacetime in which resides the mass causing those distortions, so introducing a circular, nonlinear reference into any evaluation. A condition humorously described as space telling mass where it is and mass telling space where to go!

The strange properties of gravity, combined with its relative weakness compared with other energy fields, point toward the need for a concept beyond a single spacetime dimension. Its relative weakness is not trivial. For example, two electrons repel each other electrically a million, trillion, trillion, trillion times more strongly than they attract each other gravitationally. The solution to the mystery of gravity may lie beyond the scope of our known conventional dimension set.

Relativity

Although generally considered a large-scale effect, Einstein's theory of relativity also applies to subatomic particles. We noted that the equations of relativity show that any mass accelerating toward lightspeed, as seen by a relatively stationary observer, would gain mass, slow its local time, and shorten its length along the axis of direction. At lightspeed, its mass would become infinite; time, for it, would not pass; and its length in the direction of travel would contract to zero. In practical terms, no object with any rest mass could accelerate up to lightspeed since the amount of energy required to enable that would be infinite.

Photons, the quanta of electromagnetic energy, including light, are different. Since a photon has no mass, it can and does achieve lightspeed velocity; that is its normal condition. Its limitation is that it cannot slow below the speed of light in whatever medium it is travelling. The photon is of particular interest because it always travels at lightspeed relative to any observer. Relative to itself, it is effectively travelling at an infinite velocity in a state of no time. The photon seems to experience everywhere instantaneously, throughout all spacetime, in a constant state of "now," even though we "external" observers always witness it travelling at finite lightspeed. This paradoxical condition has even led to the cynical suggestion that there might be only a single photon in the universe, appearing everywhere at once! If consciousness could instantaneously disseminate throughout time and space like that, it might support the notion of there being one uber-consciousness, sharing itself throughout our spacetime-bound experiences, and provide a hint at a solution to some of the quantum quirks we will discuss next.

MATTERS PHYSICAL

Matter is the physical characterization of reality, but our everyday notion of physical reality may be flawed. Matter is formed by mechanisms we tend to take for granted. By unquestioningly accepting common assumptions, we fall prey to the illusion of solidity and certainty. We need to unpack that assertion a little.

We become consciously aware of events about a half-second after the unconscious, but only if it considers them salient. What we believe to be solid matter is nothing but points of energy in space that feel solid because any means we have of perceiving them consists of the same virtual substance. To pursue any sense of the nature of existence, we need some appreciation of reality in terms of physics, and in particular, quantum physics. That may be an abstruse subject, but to ignore quantum physics because it is esoteric, counter-intuitive, and difficult to comprehend is to ignore a real clue about the fundamentals of our reality. Taking us closer to understanding our origins, quantum physics bridges opposing perspectives on the mysteries of existence. It embraces aspects of physics and metaphysics and provides some resonance with timeless insights intuited by ancient sages; thus, linking science and spirituality. To gain some appreciation, let us take an abbreviated look at how science describes the microscale edges of our reality.

Reductionism

"The goal of science is to know more and more about less and less until ultimately we know everything about nothing." My school physics teacher, with tongue firmly in cheek, used to quote that adage. Although meant as a joke, it contained an irony that would have resonated with the Buddha. Reductionism is a keystone of the scientific method. As we come to know more about ever-smaller subatomic events, we face the dilemma of ultimately only being able to describe our fundamental material reality in terms of building blocks of literally "no thing." At the leading edge of physics, our reality appears devoid of material content. To borrow an ancient Buddhist phrase, it is inherently empty.

That joke does seem to encapsulate the process of reductionism, though. You break the subject into ever-smaller parts and see how each works, then add them back together to get an idea of how the whole thing functions. Reductionist science has enabled spectacular progress in our comprehension, technology and sciences that would otherwise not be possible by trying to understand a complex whole in one go, as it were. It is an invaluable contributing discipline.

However, reductionism does have several drawbacks. In reconstituting complex systems from simple elements, it may not identify potentially large fluctuations that minor random variations in the initial conditions of those components could trigger in the whole. Moreover, by forever dismantling from a previously whole stage, it favours the notion, the philosophy even, of the separateness of parts, which implies that things are independent of external factors. Add to that the degree of individual specialization required to adequately investigate and understand the detailed workings of each of these ever-smaller parts, and you start to lose the ability, and even a sense of the necessity, to appropriately consider the holistic context for the complete arrangement of all those integrated constituents.

The complex whole is often much more than the sum of its parts, and thus the workings of the whole can be misconstrued and unpredictable when examined solely through the reductionist approach. This limitation is becoming recognized more in present-day scientific discussions into which the perspective of "holism" has entered more frequent usage – where an object may be more fully described as a whole interrelating with its non-separate environment, and not just as the sum of its parts. For well-balanced science, we must use both approaches.

Strings and Symmetry

One of the more novel concepts to emerge from efforts to develop a scientific theory of everything – linking quantum mechanics, gravity, and relativity – is String Theory. The String Theory model proposes matter as comprising long, thin, microscopic one-dimensional strings instead of point-like particles. The body of each string remains outside our dimension set. The vibrating ends of these strings seen from our dimension set energetically behave in different ways, which then produce the characteristic effects observed for every type of subatomic particle. These conceptual strings need more than the conventional three spatial dimensions to exist in – nine or ten, in the current thinking. These extra dimensions appear undetectable and are assumed by many string theorists to be curled up into tiny sizes.

A generalized development of String Theory by the physicist Edward Witten, postulating vibrating two-dimensional surfaces instead of one-dimensional strings, is called M-theory. One of the difficulties of string theory is that it generates an unimaginably large number of apparently workable universes with no distinct predictive qualities emerging for verification against the only universe we know. This profligate characteristic is shared with a recent extension to the Big

Bang inflation theory called chaotic inflation, which also predicts huge numbers of multiple bubble-like universes – of which ours is but one.

Symmetry is a central aspect of String Theory, and an extended conceptual form of it is "Supersymmetry" or "SUSY," which depends on a fundamental property of subatomic particles called "spin." Particle spin appears to be quantized, in that the characteristic is only expressible in whole multiples of the unit of spin. By theorizing all the possible supersymmetric combinations of particle and string characteristics, an entirely new group of as-yet-undetected complementary particles and associated forces theoretically arise, with different spin characteristics to the known incumbents of our world. Such novel, though as yet unproven supersymmetric particles would comprise opposite spin versions of each currently known particle in the Standard Model of particle physics. These are termed "sparticles," where the added 's' stands for supersymmetric.

Quantum Quirks

Quantum mechanics may be esoteric, but it is the cornerstone of existence. Although being the physics of the exceedingly small, the exceedingly small, *en masse*, is fundamental to the exceptionally large; it is the foundational physics affecting the macro world. In all but the most entrenched minds, quantum physics has transcended classical physics as the way of accurately describing the reality of atomic and subatomic behaviour. Particle and quantum physics have given us major advances in technology, and their effects extend to other fields of endeavour. In commercial terms, quantum-mechanically-related products now contribute to nearly half the global economy – so the subject is far from academic. Quantum physics gives us scientific insights into the universe and the basis of our very existence, which, contrary to widespread intuition, operates at fundamental levels on chaotic, probabilistic, and statistically aggregated bases.

Wave-Particle Duality

At the scale of quantum mechanics, nothing is certain except uncertainty. What appears to be solid is a substantial amount of space and minute presences of probability-influenced clouds of energy, just energy and force, and even force comprises exchanges of energy. For convenience, subatomic entities are still termed particles (sometimes quarticles – an abbreviation for quantum particles). That is not because they represent the solidity of even the minutest physical speck, but because it is currently still the most useful, intuitive, traditional, albeit not inclusive, shorthand analogy to use. A particle can also exhibit the characteristics of a wave. Many experiments have shown that – although to our dualistic way of thinking something could not be both a wave and a particle at the same time (that seems paradoxical) – subatomic particles can be both. As it turns out, so can entities larger than subatomic particles. Atoms, molecules, and even whole cells have exhibited this non-dual effect in the laboratory.

Conceptually, light describes multitudes of units termed photons, conceived as super-short bursts of specific electromagnetic energies and frequencies enclosed within an envelope or "wave packet" and moving at lightspeed. The apparent nature of the photon then depends on the objective of the experiment designed to detect it – that is to say, on the intention behind the way the experiment was set up to observe the photon. If the investigator intended to prove the photon a particle, then the wave packet would appear as a particle; if a wave, then it exhibited that characteristic. The evidence of what is real appears to shift depending on how we observe it. That was an extremely unsettling concept for those brought up in the safe, certain, deterministic world of Newton.

A series of international scientific meetings in Copenhagen during the late 1920s developed an interpretation for this curious situation, which has been accepted by most physicists. This critical interpretation is that photons, and by extension other subatomic particles, cannot be fully described unless included in the description are both their particle-like and their wave-like characteristics.

Uncertainty

Subatomic particles, as we understand them, appear to have surprisingly few defining attributes, chief among them: momentum (the product of velocity and mass), position in spacetime, spin (an angular rotation displacement), and polarity (or charge). Research to determine the state of a particle found that one could never precisely pin down specific complementary pairs of characteristics simultaneously. Werner Karl Heisenberg, a leading voice in quantum physics, developed the "uncertainty principle," which states that both of a pair of such attributes cannot be completely known. The more you know about one, the less you know about the other. Taking velocity and position as an example, if one were to determine a particle's speed, one could not accurately know its location, or if the location was precisely known, its velocity was not. Complete knowledge of a particle's state can never be established.

At first, this phenomenon was blamed on inadequate measurement techniques, though now it is recognized that this uncertainty is an inherent characteristic of the particle. Such uncertainty appeared to be in the particle's nature and a fundamental characteristic of all subatomic particles. That means that a particle can never be stationary, for if it were, then both its location and velocity (zero) would be known, which conflicts with the uncertainty principle. Indeed, particles are always active, vibrating, their complete state forever uncertain – they are said to exhibit the jitters. Although a measure of that uncertainty may be calculable, probability has been introduced front and centre in seeking a description of the fundamental building blocks of the universe.

Many types of subatomic particles form the central components of atoms. These comprise the nucleus, made up of protons and often neutrons, which in turn comprise triplets of quarks and surrounding orbits of electrons. The electrons forming an atom are more like surrounding shells of probabilistic energy clouds, or concentric bubbles within bubbles, where the closest that one can come to defining their absolute state is "likely in the vicinity." Even the quarks – associated with subatomic nuclear and electromagnetic forces that attract and repel various particles, thereby holding the atom together while also preventing it from collapsing – exist as a continuous jittery turmoil of energy-exchanging forces and may even include transient antimatter versions of themselves. Solidity is nothing but jittery exchanges of energy.

Superposition

When a particle is emitted from a subatomic event or source, it cannot be considered as an actual particle until an observation or other interaction evidences it. We will discuss this aspect shortly. Until that detection occurs, it can only be considered as a possible nascent particle or a particle-to-be. For simplicity, i have chosen to distinguish an emitted but undetected particle by the term "pre-particle."

As experimentation technology improved, the fledgling theory of quantum mechanics was not only born but proven effective. Sophisticated experiments showed that if detecting which of two paths a single quantum pre-particle might take, it occupied both until observed. Only at observation would it appear on one path or the other. The conclusion was that before detection, the pre-particle must have been on both paths at the same time.

Unsurprisingly, this controversial phenomenon was investigated rigorously but repeatedly demonstrated the same counter-intuitive result. The pre-particle spreads its options by taking both paths until observed, after which it settles for being observed as a classical particle on one path or the other. The nature of this settling was probabilistic. The clear experimental evidence was not that the particle had a 50% probability of being on one path or the other, but that it split its existence in two, and each split part had a 50% probability of being found wholly on that path and not the other. Before observation, the pre-particle's location was in neither one place nor the other; it was simultaneously in both places.

The unobserved pre-particle's partial presence at one location appeared superimposed onto its quasi-presence at the other, to represent the whole particle at observation. This ability to be in more than one place at the same time – unless observed – was termed "superposition." Further evidence extended this condition of superposition beyond two alternate paths to being in any number of simultaneous, probability-weighted possible states until the particle is detected as being in a unique classical state.

A pre-particle in superposition is not in any specific state; its full range of potential – all possible states – is available to it. Each possibility is probability-weighted as to its chance of being in the actual manifest state when determined by observation. That probability weighting may be dynamic – influenceable by external factors; more on that possibility later. Regardless, all possibilities exist within the pre-particle's non-physical condition. All superpositioned states are potentially at the point of "happening" at once. In this Alice-in-Wonderland existence, our level of consciousness would not be able to cope. We could not navigate such a world. We might even speculate that our experienced direct relationship between cause and effect might be a special case; one of existential benefit to this universe's more linearly operating sentient participants.

Wavefunction

The schizophrenic nature of a pre-particle, given multiple alternatives of potential paths or states in superposition, needed a formula to accommodate all possibilities. Erwin Schrödinger conceived the quantum particle's pre-emergent state of existence as being within an envelope of all possible outcomes that he delineated by a mathematical equation called a wavefunction. The wavefunction equation highlights the probabilistic nature of the pre-particle by describing all its possible states while it is undetected and yields the associated probabilities for the particle manifesting in any possible state. While the wavefunction envelope remains intact, it describes a pre-particle possessing any set of all its possible characteristics in superposition with one another. These include its location, momentum, and spin; collectively called its state.

Schrödinger concluded that a pre-particle in superposition within a wavefunction was not in any real physical state. It did not materially exist in spacetime until intruded on by an intentional observation. What travels the universe is the wavefunction of the pre-particle, containing all possible superpositioned potential states of it and their respective likelihoods – not an actual particle at all. An act of conscious observation impinges on the possibility-wavefunction and extracts the pre-particle's imminent probability-influenced manifest state information for the observer, triggering a "collapse" into that specific state. The surrendering of that status information by the pre-particle in superposition causes it to behave like a classical particle. It becomes "real."

Physicists may use "decohere" as an alternative term to "collapse." This term considers the spectrum of every possible state of the pre-particle in superposition as being in a "coherent" state of mutual resonance within the wavefunction. An act of observation destabilizes that coherence, causing all the resonant possibilities to "decohere," to fall away, leaving the probabilistic choice to emerge. In the speculative "Many-worlds" theory, all decohered possibilities were thought to survive and materialize in inaccessible alternate realities. The term may be technically superior, though i find "collapse" to be a more accessible metaphor to conjure. I can just imagine a wavefunction floating around spacetime like some enormous, undulating soap bubble of probable pre-particle possibilities, then collapsing into the residue of a now-certain state of being a classical particle because of the pinprick of an information-extracting observation.

Some experiments suggest that the manner and intensity of conscious observation – the intent of the observer – may have some influence on the outcome of the pre-particle's state as its wavefunction collapses. That effect may be so, even if the particle-conscious observer is remote from the locality of the particle's manifestation. This apparent link has led to the controversial conclusion that sentient consciousness might influence the probable outcome of a particle's wavefunction collapse in a manner unconstrained by spacetime limitations. In the presence of such an effect, the scientific ideal of a purely objective observer of an experiment may not be feasible.

Quantum Jitters

The creation, maintenance, and evolution of existence require continuous jittery exchanges of energy. Any material form can only result from continuous subatomic change. Any function or process involves changes, and the notion of time appears to be one of experiencing sequential change. The foundation of our entire observed reality is built upon sequential snapshots of the historical summation of probability-influenced events born of subatomic, statistically aggregated, quantum level changes that we are calling quantum jitters. Material reality appears as a complex, feverish juggling act, which cannot stop or the illusion of a substantial arc of physical form falls apart; its apparent solidity dissolves. Existence is change and change is existence. That mirrors the Buddhist aphorism – form is emptiness and emptiness is form – for, without energetic change, the reality of form must collapse back to whence it came, into its initial state of emptiness – nothing – because only nothing requires no change.

Now we can appreciate the interdependent, holistic nature of the structure of our reality. Matter and force are but energy existing in a flux of change. Gravity is spacetime warped by mass and energy. Time is the interval between local changes and is influenced by matter density and relative speed. Not only are these basic features of existence correlated, but they also appear interconnected, networked, like threads in a tapestry. Below the surface of reality, all these attributes are virtually equivalent. At the subatomic level – our basement of perceived reality – the illusion of solidity is maintained universally by virtual energetic quantum jitters.

However, this essential quantum jittering might conjure up the idea of a constant agitation of all particles in the universe, which seems extravagantly energy intensive. Perhaps a more economical alternative perspective on maintaining form might be conceived involving less energy. A general principle in data compression is to reduce the volume of data by focusing primarily on those elements that provide new information through change. Extending that data-compression principle to the universe at large, one could speculate that only pre-particles providing new information for sensing by sentience – those required to participate in the manifestation of observed form – need have their quantum states refreshed: only *they* need to jitter.

A conscious mind processes information and confers meaning. The mind extracts and processes externally-derived information via the various senses and sensors at its disposal, or by generating it within the mind – thought being one of those senses. Any prospective quantum pre-particles not entangled with sentience – those not required to provide reality-rendering information for any form of consciousness – could remain in their pre-manifest quantum potential state of superposition, with no definable characteristics. Thus, those pre-particles – the vast majority – would not require energy to jitter, because they remain unrealized possibilities in a state of superposition within their wavefunction. The presence of sentience would only be nurturing energy-induced form within its limited observable environment.

Whatever the total size of the universe is, at its increasing rate of expansion, our earth-centric sensors can observe less and less of it. Therefore, most potential pre-particles in superposition beyond our observable horizon would never have their wavefunction collapse. They would not need to jitter – not on our account anyway. They need never leave their wavefunction-bounded state of uncertain grace of pure potential. Immaterial pre-particles represent the uncalled understudy waiting in the wings, requiring little or no energy-intensive participation, unless called upon by sentient awareness. Since humans can perceive only a tiny fraction of this trillion-trillion-starred universe, and only a fraction of that is recognized as atomic matter, it seems reasonable to infer that the impact of Earth-bound sentience on the need for the universe to maintain an energy-intensive material form would be minimal.

Virtual Particles

A curious effect of the jittery nature of subatomic particles is that they can wink in and out of existence spontaneously in a vacuum-energy field existing throughout spacetime. Termed virtual particle-pairs, they continuously occur naturally, generated spontaneously throughout the universe. Although this phenomenon is a well-established quantum process, it still has little in the way of any real explanation. Due to the energy potential trapped in the vacuum of space and the jittery nature of pre-particles, virtual pairs of quantum particles seem to have a significant probability of squeezing out of the vacuum. These cause a constant flux of newly produced quantum particles, and their corresponding anti-particles, to appear. Generally, one particle of the pair and its adjacent antimatter twin quickly find each other and mutually annihilate, radiating a photon in the process, thus almost neutralizing their brief appearance. However, that new photon does increase the background radiant energy of the universe.

Some of these virtual particle pairs do not self-annihilate, however. Any local intense gravitational or electromagnetic field that happens to be where the virtual particle pairs first appear could cause the pair to split up, and for each to take a different route through spacetime. The stray anti-particle of such a split couple could eventually meet up with an ordinary particle and annihilate it, leaving nothing but a photon. Alternatively, it might become sidelined by attraction into a black hole's gravitational field. Black holes act as mass-energy sinks; that is to say, everything entering them, including light, becomes unavailable to the rest of the universe, except for the gravitational force resulting from the mass of the black hole.

However, when interacting with the anti-particle member of a virtual particle pair in this way, black holes would allow the matter-compatible member of the pair to remain permanently in the universe, thus appearing to act as a matter-generator. Only apparently, however, since the antimatter particle entering the black hole should reduce the black hole's mass by annihilating an incumbent tenant particle. In essence, the black hole and its massive gravity would shrink a little. The net effect would be that, in the absence of new fuel, the black hole appears to evaporate. This phenomenon is known as Hawking radiation, after Stephen Hawking, who conceived it. Thus, whatever happens to a virtual particle pair, it seems that some lasting effect remains within the universe in terms of either a residual increment of particle mass or photonic energy.

The constant materializing and dematerializing of virtual particles imply that there might be an extradimensional field component involved in their spontaneous appearance. The vacuum-energy of space could represent an interface with an energetic extradimensional field, the source of these largely self-neutralizing quantum events. Such an interface generating continuous quantum-state particles and antiparticles could interact with the physical particles and hence with the structure of matter as a whole. It might even modulate the form of the bubble of our spacetime existence – the universe. As an extradimensional field, it could supply or extract energy and be in contact with our manifest universe at all points of its existence. Furthermore, such a field may only appear to saturate the whole universe because the universe exists in it, not the other way around.

The first law of thermodynamics stipulates the conservation of energy in a closed system. However, in the presence of any such extradimensional interface, the universe might not be an entirely closed system. Effectively, it might be slightly porous, and so whichever path these emerging virtual particles take – adding particle mass or photonic energy – since mass and energy are effectively interchangeable, the total energy of the whole universe would incrementally increase. Our slightly porous bubble-like universe might be pressurized with additional mass-energy. With no counter-pressure on the "outside" boundary of our spacetime to resist such expansion – since there is no outside – the universal expansion rate simply accelerates. This virtual particle-pair pressurizing effect may even relate to the so-called cosmological constant, which, under the more popular title of "Dark Energy," is the current prime candidate for causing the observed accelerating expansion of our universe.

Entanglement

When two quantum pre-particles proceed from the same event, they emerge in a state of mutual "quantum entanglement." Entanglement means that the state of each pre-particle can only be fully described by reference to that of the other. The state of either is unknowable until observed; however, if one entangled pre-particle's wavefunction were to be collapsed by an observation, the other's state will always correlate with it, even at a great distance apart. The remote entangled particle's state is thus deducible before any confirming observation.

Nonlocality

Many experiments have verified this mutual-correlation effect of entangled pre-particle wavefunctions. These have also shown that the collapse of the remote second particle into a state that invariably correlates with that of the first can occur at significant distances, before any lightspeed communication between them could be possible. Such an apparent impossibility has no rigorous explanation and is known as the quantum enigma. This remarkable effect demonstrating apparent simultaneous information-sharing between entangled pairs of quantum pre-particles at distances greater than information could travel at lightspeed is termed the principle of "nonlocality."

The theoretical physicist, John Wheeler, had proposed a thought experiment intended to clarify the temporal nature of such quantum effects. However, two decades were to pass before his hypothesis could be confirmed in the laboratory by an experiment, somewhat dramatically called the "Delayed Choice Quantum Eraser." The outcome of this experiment showed that a full description of entangled particle pairs when the test begins could only be determined by things that do not happen until the experiment finishes. In other words, entangled quantum events appear able to influence their historical origins!

These findings further confounded the quantum enigma, extending it to include evidence that correlation among entangled pre-particles may occur across great distances not just instantly, but independently of time. The experience of one pre-particle appears capable of affecting the state of its entangled sibling retroactively, implying that a quantum effect can precede its cause. So, the principle of nonlocality applies not just spatially, but also temporally. At the subatomic scale of things, it appears that lightspeed and the arrow of time are not the immutable barriers they represent to us. Wheeler pointed out that if these effects were applied to a similar apparatus of interstellar proportions (and the universe could constitute such an apparatus) a last-minute decision on Earth about how to observe a photon from a distant star could determine the nature of that photon's origination reality billions of years ago.

This phenomenon leads to the unsettling conclusion that from our linear spacetime perspective, the cause that triggered the collapse of an entangled pre-particle from superposition could occur *after* the wavefunction-collapsing event. This realization led to a flurry of research aimed at pursuing the related possibility of faster-than-light communication using entanglement and nonlocal effects. However, although breaking the lightspeed barrier remains a dream of many physicists, it seems that nature may have some behind-the-scenes access to abilities unavailable to us.

Nonlocality is a counter-intuitive new reality that Einstein described as "spooky." Nevertheless, it turns out to be a critically important factor in comprehending existence. Not only might the correlation and realization of pre-particles be independent of conventional spacetime, but also triggering a particle manifestation may not have to precede the experience of the pre-particle being manifested. Thus, the extraordinary conclusion considered here is that at this fundamental level of existence, influencing a past creation event and its descriptive history leading up to the present could occur after the fact. The nonlocal entanglement phenomenon, in which a continuous state of mutually informed correlation exists independent of spacetime restrictions, has profound implications in the context of our understanding of the nature of existence.

Furthermore, the question arises as to how such phenomena might pertain to attributes of a metaphysical nature – less obviously connected to quantum mechanics, but more related to consciousness and sentience. This begs the pivotal question of whether nonconscious processes might also be nonlocal phenomena, unconstrained by the dimensions of spacetime and capable of entanglement with superpositioned elements leading to events that for us are already in the past. When viewed from within our limiting perspective of linear spacetime, present-day nonconscious awareness might entangle with historical pre-particle realization and influence their manifestation retroactively.

Quantum Concepts Takeaway

This brief, intense, and necessarily superficial understanding of the quantum world attempts to establish some passing familiarity with aspects derived from quantum physics – particularly the counterintuitive ones. Enough, at least, to appreciate the implication that reality is not what it seems, and not based on the Newtonian paradigm of the last few centuries, with its themes of clockwork-like mechanisms, billiard ball-like atoms, and the confident certainty of determinism. Most importantly, in this context, is the inference by some aspects of quantum physics that physical reality might be inexorably linked with participating consciousness. Both consciousness and quantum mechanics may share some similar mysterious qualities, which, when better understood, may establish a fundamental link between them. Indeed, consciousness may prove to be associated with, utilize, cause, catalyze, or influence quantum phenomena.

That challenge to our intuitive understanding of reality may introduce the role of consciousness as a fundamental process of existence. That is known as the hard problem of consciousness, and when linked with quantum mechanics is among the most tentatively understood phenomena with which science has ever dealt. So contrary is it for theoretical physicists to have to consider the implications of consciousness that often the preference is to ignore it altogether. Indeed, this uncomfortable but stubbornly persistent link of physics with consciousness is referred to as the "physicist's skeleton in the closet."

If these counterintuitive effects of quantum physics elude your understanding, you are in great company, including many pre-eminent contemporary physicists – Einstein among them. Nevertheless, even though they may be bewildering, these weird quantum effects have been experimentally proven repeatedly. Quantum behaviour, though poorly understood, is a characteristic embedded in the actual nature of things and not an arbitrary human description. However, though the effects are proven, that does not mean the mechanisms behind them are understood. This chapter has attempted to provide some appreciation for those provocative quantum anomalies that form quantum theory – one of the most tried and trusted scientific theories of all time.

Now, on quite a different scale, we will look at how the whole universe may have started.

CONTEMPLATING ORIGINS

How did physical form on a cosmological scale come to be – the beginning of the universe, our existential origin? Contemporary science currently proposes several ideas that mostly agree with the observations our instruments have recorded; however, they all remain incomplete theories. Some skeptical scientists assert that the central theory discussed here, the Big Bang Theory, is still a highly speculative explanation and an unnecessarily complicated one at that. It requires many synchronistic, highly improbable, incredibly precise random coincidences to happen, to get what we observe today.

Other explanations pondered in scientific circles may address certain observed phenomena better, though that does not make them right either. Each one appears to have countervailing flaws in some other area that reduces their competitive value as a potential usurping theory. If they had not, then everyone would agree on that new theory. That is how science works; the very latest, best, and often the simplest explanation of what is going on is usually the most acceptable as a working theory until new evidence comes along to prove it wrong or enhance it, thus establishing a more complete theory. In science, it is axiomatic that no theory is proven, but is always open to improvement or disproof.

Many brilliant scientists are working on different models and interpretations for cosmology, quantum gravity, and many other unsolved aspects of theoretical physics. No single theory works yet, so this field of endeavour remains wide open. After reviewing some mainstream scientific viewpoints in this chapter, i will offer a few additional thoughts of my own. I am not qualified to offer these as theories, however, some of these speculative ideas arose during contemplative meditation, so i have tried to interpret them within my limited understanding of present-day scientific views.

Expanding Universe

The present-day cosmological view of the formation of the universe holds that the origin was a quantum singularity of virtually infinite energy that just popped into existence. It appeared from nothing and started a brief period of extremely rapid inflation, which then morphed into a more uniformly paced expansion. The popular term for this explosive explanation is the Big Bang Theory.

The term "singularity" is physics-speak for a dot, smaller than a dot really – a point so tiny as to be effectively dimensionless. Indeed, until this hypothetical point started inflating in size, there were no dimensions, so it effectively created them as it went along. A dimensionless point remains just a concept, necessitated by our inability to imagine a state of non-dimension – no size, and possibly no time. If the singularity credited with starting the universe seems an anomaly to present physical laws, it is perhaps due to our inability to conceptualize it – being limited by lifelong exposure to the current physics governed by our familiar dimension set. On the other hand, it might just be wrong.

When we play the dimensionally burdened Big Bang movie of our expanding universe backwards, we appear to arrive at a virtual beginning point of spacetime as being zero – the singularity. We reach a stage when we can no longer imagine anything that preceded it. One is expected to simply accept the implausibility of an infinite amount of energy emanating from a dimensionless point before time began. Such acceptance might be essentially an act of faith. Despite the mathematical models that we contrive, we cannot practically grasp an energy-driven creation of matter from nothing, coupled with the unfolding of dimensions – particularly the so-called dimension of time. It seems as paradoxical as lifting yourself by your bootstraps; we will revisit that shortly.

Nevertheless, the theory goes, something came from nothing. Pure energy was condensing into form with unfurling spacetime dimensions to guide it, like a red carpet unrolling in front of honoured guests. These dimensions allowed it to change, rapidly expand, and then evolve. For a questionably brief period, estimated to last some 10^{-43} seconds, this expansion is posited to have accelerated exponentially and is described technically as inflation. The dense, hot, opaque turmoil of energetic "pre-stuff" – possibly a quark-gluon plasma – created during this vigorous initial inflationary expansion phase, prevented any atomic matter from forming, or any electromagnetic radiation, including light, from travelling through it. Therefore, within this inflating kernel – this proto-universe – no mechanical or radiative communication was possible among its unformed elementary constituents that were rapidly dispersing across the inflating space. Thus, the formative primordial plasma soup was effectively homogeneous, and coalescence of any potential matter could not occur during this inflationary phase.

A variation on this simple inflation theory, called chaotic inflation, essentially proposes that quantum effects occurred within those initial conditions of this as-yet embryonic universe. These effects may have caused random density variations within the primordial plasma soup, rendering it something less than uniformly homogenous. At such an early stage of formation, such random variations then developed into significantly non-uniform, patchy physical characteristics within an expanding nascent universe. In other words, different parts of the universe-to-be might have developed different mass densities, or even, perhaps, different physics. As a result, this energetic embryo could have condensed into multiple clusters made up of like forms of matter, or into zones of such different physics as to effectively become multiple segregated universes, of which one turns out to be ours.

Interestingly, this concept of chaotic inflation, like the previously mentioned String Theory, also postulates the possibility of multiple universes. We are not talking just a few, either. Rough order-of-magnitude estimates project more than 10^{500} possible versions of them. That is a preposterously large number, one followed by 500 zeros, which for all practical purposes, may as well be infinite. Cosmological theories discussed here only apply to the development of the physics in our observable part of any such universe or multiverse.

After that brief inflationary period in the only universe we know, at an estimated 10^{-36} seconds from the origin – still a ridiculously short period – a reduction in energy density resulting from the expanding space containing it took place. That energy density reduction caused temperature and pressure reductions, and possibly several phase transitions. These may have led to an asymmetric decoupling of fundamental newborn forces from their theoretical "parent" – an assumed single primordial unifying force – that effectively exhausted the inflationary part of the expansion. Each force decoupling may have occurred at slightly different times, perhaps during different phase transitions of the inflating plasma soup, so isolating the four fundamental forces comprising electromagnetism, the strong and weak nuclear forces, and the so-called "force" of gravity.

Further cooling of the universe by continued expansion and differentiating of these forces over the next three minutes led to the beginning of nucleation of the proto-matter plasma soup into pre-atomic elements, such as electrons and a variety of quarks. The probability of everything coming together, in such a finely tuned state at random is estimated to be around one chance in a trillion times itself ten times over – long odds indeed.

Continued expansion and cooling of the universe's opaque energy, over the next three or four hundred thousand years – note the dramatic change in time scale – led to electrons and quark-formed proton and neutron nuclei coalescing into the first, simplest and lightest, atomic structures, such as those of hydrogen, deuterium, helium, and lesser amounts of lithium. This condensation process enabled space to clarify into relatively low-density semi-transparency, which allowed the previously trapped electromagnetic radiation spectrum, including light and heat, to propagate across the new spacetime. The initial chaotic-inflation-inspired variations in density and subsequent increase of gravitational effects caused the accretion of the newly formed atomic material into clusters. These hotspots of density fluctuations developed local gravitational fields, which then accelerated the formation of larger clumps of matter, further amplifying the gravitational effects. This gravitationally driven accretion process created the first stars and galaxies throughout the next few hundred million years.

Stars form through the process of gravitational accumulation of large amounts of nearby material, mostly gas, which then become compressed under its own weight to such an extent as to initiate atomic fusion, thereby giving off copious amounts of radiation. The effects of the star's nuclear fusion power form the atoms of the elements carbon, nitrogen, oxygen, and all the way through the Periodic Table to iron. The remarkably unlikely creation of carbon, essential for later carbon-based life forms like us, is itself another miracle of fine-tuning. The only way we can imagine it happening is by an impossibly fortuitous coincidence of a simultaneous triple collision of quantum-state helium nuclei particles. These triplets, at just the right conditions of fusion temperature and pressure of their surroundings within the star, could have formed stable carbon atom nuclei. This seems a highly improbable outcome – though, fortunately for us, an outcome, nonetheless.

Expiring Universe

As the Buddha observed, "everything that arises must cease," and right at the other end of the scale of existence, scientists have projected several scenarios for the eventual demise of our universe, billions of years hence. These scenarios have completely reversed themselves in just the last few decades. Until mid-last century, conventional wisdom, that of Einstein included, considered the universe as being static – the "Steady State" theory – with neither expansion nor contraction occurring.

Subsequently, after Hubble's observations of receding star systems showed the universe to be expanding, it was thought to be headed toward a gradual slowing-down of expansion until it would stop. After stopping, it would then commence contracting back from whence it came, into a tiny, yet massively energetic bubble. This theory – described in jocular terms as the "Big Crunch" – is no longer regarded as a valid interpretation, since there is no evidence for any future slowing of the current expansion rate. On the contrary, there appears to be a puzzling acceleration of the expansion rate of the universe in just the last few billion years that has no adequate explanation.

Based on this recently observed accelerating expansion, the latest cosmological theory for the end of the Universe – dubbed the "Big Rip" – projects an essentially energy-exhausted, minimal density, cold universe of vast disintegrating nothingness – maximal entropy. The conflicting nature of these quite recent prognostications does little to inspire confidence in their quality.

Dark Universe

Current cosmological estimates of the physical content of the universe show that there is not nearly enough detectable mass-energy in the known universe to account for its accelerating expansion and galactic formation behaviour. The shortfall is not trivial. Only a small percent of what is necessary has been detected or estimated to exist as ordinary physical atomic matter. That led to the speculation that there must be much more matter or mass-energy-equivalent around. However, it must be of such a nature that renders it undetectable directly by our instruments. Thus, the term "Dark Matter" arose to explain this missing mass. This dark matter would have to be such a mysterious unknown form that, though not observable – not reflecting or emitting electromagnetic radiation – it could somehow create the required mass-like gravitational effects to hold the galaxies together in the expanding universe.

The unobservable presence of the gravitational effects caused by dark matter has been inferred indirectly by calculating theoretical galaxy spin rates and comparing them with their apparent mass distribution. The difference between the gravity resulting from the mass distribution of the galaxy, and the gravity needed to hold the spinning galaxy masses together represents gravity associated with dark matter. An alternative detection method, called gravitational lensing, results when light coming toward us from a very bright distant source such as a quasar passes through spacetime that has been distorted by an intervening source of gravity. This distortion effectively bends the path of the light. The source of light bending might be an intermediate galaxy, massive neutron star, black hole, or undetermined mass-like gravity such as dark matter. If the amount of light-bending is not fully explained by the number of observed star systems in that area, and all other known gravitational sources in the intervening area are eliminated, then what remains is considered as being dark matter. The technique is so named because the resulting bending of the light path through spacetime appears as if a giant lens were placed between the distant light source and us.

As if this unknown source of mass-like gravity were not enough of a mystery, the accelerating aspect of the universe's expansion can only be explained currently by the inclusion of a small cosmological constant into Einstein's formula of relativity. That constant implies the existence of a large amount of so-far undetectable energy, causing a mysterious repulsive force, which acts somewhat like a negative form of gravity, pulling apart larger masses such as galaxies, effectively expanding the universe. The popular term for this effect is "Dark Energy." Again, though speculative, its impact is measurable and is extremely significant. Theoretically, this dark energy provides a large fraction of the total mass-energy forces estimated as needed to match the observed accelerating expansion rate of the universe.

Dark energy, when combined with dark matter, represents over 95% of the universe's total mass-energy calculated as necessary for it to behave the way it does.

Although there are conflicting accounts for the origin and demise of the universe, it is generally acknowledged that it has been expanding from some form of small energetic kernel for nearly 14 billion years. Within that timescale, our Sun, the planet Earth, and the rest of the solar system all formed four to five billion years ago.

One of the frustrations of cosmology is that estimates and observations about the universe are limited to the observable part of it. This limitation is due to the expansion of the universe, whereby far distant quasars and galaxy constellations are accelerating away from us so fast that they just disappear from view. As a result of the finite speed of light coming toward us from them, coupled with the ever-increasing expansion rate of the space receding away from us, we are prevented from observing anything of the universe that might lie beyond a roughly 46 billion light-year "horizon."

Thus, the observable universe in which we live approximates to a 93 billion light-year diameter spacetime bubble and anything postulated for existing beyond that horizon is entirely speculative. We simply do not know if our observable universe represents a small or large fractional portion of a larger universe or multiverse.

Other Notions

There is a lot we do not know. Much of what we think we know is scientific speculation and the current mainstream community of scientific opinion may not unanimously accept them. Many reservations and alternative notions are pondered, of which a selection, including some of my thoughts, briefly follow.

The Singularity

Perhaps the most hotly debated speculation in cosmology centres around how the universe could have started. It is all very well proposing the origin as a hot quantum singularity emerging spontaneously out of nothing, but that seems to omit quite a bit of detail. Obvious questions centre on how that event occurred – to say it just happened sounds akin to believing in magic – and what was happening just beforehand – the assertion that time had not started yet, and therefore nothing preceded it, is unsatisfactory. There may never be the opportunity to probe up to, into, and before that first moment. Nevertheless, it is incredible just how far back in time we can see with our powerful telescopes.

Still, we may never be able to see much past that turmoil of opaque plasma that characterized the initial phases of the proto universe. Yet, never is a long time, and current work into cosmic background radiation and gravitational distortion echoes may provide more clues. Meanwhile, reasonable theories have developed that seem mainly cogent and lend some credence to otherwise unfalsifiable explanations. The Big Bang singularity is one of them.

Non-linear Time

An alternative view about time questions whether a dimensionless and timeless singularity could have originated both time and space simultaneously. It suggests that perhaps such portrayed beginnings might merely be the invention of human linear thinking. Running the expanding universe "movie" linearly backwards in time to arrive at a year "dot" origin may be misleading, for it assumes time as being linear.

However, if you were to imagine time as a non-linear continuum of sequential changes, particularly at the earliest originating moments of the universe, or even pre-existent, you could envisage a universe starting not at time zero but partway along a nonlinear conventional temporal axis. That does not resolve the issue of the improbable appearance of a small, finite-sized kernel of massive energy, though it does avoid the temporal anomalies of a singularity and perhaps the credibility issues of the rapid inflationary phase. Such a gentler alternative beginning to the universe might be metaphorically characterized more as a flower blooming than a big bang booming.

Cyclic Universe

Another view, questioning the temporal aspects of the singularity, is the cyclic universe formation. Essentially, the thought is that the universe repeats cycles of growth and collapse along a temporal axis – oscillating in and out of existence. Each new cycle assumes a complete restart, with no memory of information from a prior existence. This concept worked better when the idea of the day was that the universe would eventually stop expanding and contract back to zero – the Big Crunch. However, it has been resurrected in a modified form to suit the Big Rip scenario by assuming that state phase transitions occur in the ever-expanding, low-density universe, resulting in one or more new universes being born from the disintegrated remnants.

Multi-universe

The multi-universe concept also allows for a time continuum being already in place, during which new universes bubble into existence. This approach, in which time operates before each new universe's origin, seems more appealing, though it does not necessarily reject the concept of time as a dimension. However, time does not have to be a dimension at all but just represents sequenced change, and so is ubiquitous and limitless.

Time before Origin

Another argument challenging time as being a dimension that started at the big bang is that if a spatial singularity origin occurred causing a new universe to form, there would have had to be a period when it was non-existent, for any phase-change of state, quantum or otherwise, to then occur, and bring it into existence. Thus, some form of temporal pre-existence to allow for such a change would have to precede the singularity event to accommodate the change-introduction event itself. Time could not originate from within that singularity, or else there could not have been one.

Membrane Origin

In terms of understanding the universe's origin, a membrane-based conjecture appears to be based on recent developments of the extradimensional, supersymmetric string theory. We currently experience spacetime as a volumetric dimension coupled with a time-like component. In grossly simplistic terms, a speculative event based on M-theory – an over-arching form of super-string theory – postulates that our volumetric spacetime existence could act as a "surface" of an extradimensional membrane, like a two-dimensional patch existing on the surface of a three-dimensional balloon.

This variation for the universe's origin proposes that two such extradimensional membranes collided, created a massive quantum fluctuation at the impact site – the singularity of the origin. This massive energetic quantum fluctuation initiated the creation and expansion phases of the spacetime-dimensioned universe along Big Bang lines. This alternative explanation to the inflating singularity phase of the Big Bang is called the Ekpyrotic universe – born of fire. It avoids the temporal and spatial singularity and flatness issues of inflation.

Personal Thoughts

A Binary Universe

One of the more unsettling aspects of the quantum singularity concept in the Big Bang theory is the paradoxical aspect of something coming from nothing. However, after contemplating this conundrum for a while, it occurred to me that there is a relatively simple analogy of how something might arise from nothing – *ex nihilo*. In school mathematics, when simplifying terms on both sides of an algebraic equation, you might remember having sometimes erroneously "proved" that zero equals zero! No? Well, perhaps it was just me.

This trivial analogy is a bit like that. It may sound like mathematical sleight-of-hand, but, at the quantum level, real matter originates from nothing all the time in space – hence the designations of virtual particle-pairs – so this idea is symbolically trying to describe an already "real" process. The crux of this thought is that matter produced by vacuum energy – manifesting as pairs of virtual particles – usually does not last because the particle and anti-particle products quickly mutually annihilate. However, some can continue on their own, provided the symmetry of their formative relationship is broken after formation but *before* they can mutually annihilate.

At its mathematical simplest, one added to its opposite, minus one, produces zero – nothing. By reversing the written order of that equation, two real numbers, one and its opposite, minus one, can ensue from zero, or, like the virtual particles, so can a set of anything and a set of anti-anything. Thus, at the quantum level, something can quite conceivably emerge from nothing provided its antithesis also emerges. If the antithesis were then to become segregated from its partner, the symmetrical formative relationship would be broken, preventing the entity and anti-entity from annihilating each other, thus allowing them to remain independent.

From this perspective, given the balancing nature of something and its antithesis, the productive power of nothing is infinite. So, anything could result from nothing, provided the symmetry of its formative pair is broken in some manner. Otherwise, the anything and anti-anything annihilate each other at the outset. Still, this is just a simple example, and we are talking about entire universes here. Could this principle extend to such assumed complexity?

By extending this *ex-nihilo* thought, an alternative variation to the Big Bang theory arises. A deep primordial vacuum could have created an unimaginably intense high-energy quantum virtual particle-pair – enough to kick-start the first link in the quantum kernel and universe-creation chain. Each one of the pair might usually annihilate the other, returning everything just created to nothing – net sum zero. However, if a symmetry-breaking quantum event were to occur concurrently with the virtual particle-pair emission, it could inhibit the virtual particles from self-annihilating back to nothing.

The idea of a binary universe stemmed from a combination of the *ex-nihilo* concept and insights developed from frequent contemplative enquiries into the meaning behind the recurring meditative motif that i described at the end of Part 1. The central theme of that motif consisted of an individual pulling two face-to-face planks violently apart, generating a dazzling pair of sparks from within the resulting gap, and then twisting one plank through a quarter-turn relative to the other. The orthogonal twisting of those surfaces was expressly significant and appeared to allow the pair of sparks to remain.

The significance of this meditative image lies in what appeared to be happening, not how. My experience as a kid was that if you smacked two flat planks against each other just so, the brief surge of high-pressure air escaping from between the planks could create a satisfying crack of sound so sharp it sounded like a gunshot. If this effect were physically possible to reverse, it might create a momentary extreme low-pressure zone between the separating plank faces. Perhaps, at the quantum scale, if two membrane surfaces separated fast enough, the resulting super-vacuum effect could act as a temporary virtual particle generator.

Although the spontaneous manifestation of quantum virtual particle-pairs is continually occurring in the vacuum of space, at the universe's origin there would be no distracting black holes around to potentially sideline one of the virtual particles since the universe did not exist yet. The massively energetic pair of quantum virtual particles in the plank metaphor would have to be restrained in some other way to prevent immediate mutual annihilation. That seems the sole message in the final orthogonal twisting of the metaphoric planks. It symbolically broke the symmetry of the virtual particle-pair formation process, preventing their annihilation.

In contemplating the possible mechanics of such symmetry breaking, one possibility that arose was that since each virtual particle-pair kernel is subject to quantum jitters, each one might acquire its own unique set of initial conditions as its initial primeval quantum wavefunctions collapsed. In a dimensionless setting, with no external constraints, such initial conditions might result in nonaligned attributes that include different dimensional structures for each kernel as they start their rapid expansion from nothing.

Such different dimension sets might be incompatible enough to prevent their mutual interaction, breaking the symmetric relationship of the pair and thus forestalling annihilation. So, even if each elementary energy kernel were to remain co-located, since their dimension sets would be incompatible there would still be no physical interaction. Each particle could then inflate within its unique dimension set, independent of the other. The two could develop separate identities, despite originating as antithetical opposites. They could even remain superimposed on each other since they could still not directly interact and therefore could not return their net energy to zero through mutual annihilation.

Mutual Development

Of these two independently developing embryonic proto-universes, one would be "ours" – the other would not. One member of this binary-universe system develops the way "ours" did – the other probably differently. "Our" universe expands more or less along the lines of the latest cosmological theory outlined earlier. We cannot know what sort of physics or dimension set would have developed in the anti-universe. However, a similar process of causing mass, if replicated to any degree within the anti-universe twin, might create a gravitational linkage between the two co-located universes.

Such an extradimensional tidal-like linkage could synchronize expansion rates and even galactic formations among the twin universes. On a cosmological scale, that synchronization would tend to eventually cause the coinciding of concentrated mass distributions among the otherwise segregated though superimposed worlds, particularly among those more massive entities with which we are familiar, such as black holes and galaxies.

That this anti-universe may have formed initially from an anti-particle should not imply that it consists of what we know as supersymmetric antipartners. Since the term "anti-universe" might encourage such assumptions, we can instead refer to this hidden sibling as the "shadow" universe. The term "shadow," borrowed from psychology, evokes the intangible nature of a non-dual unconscious relationship as might exist within sentient entities. It also underscores the inseparability and interdependence of these two diverse worlds – you cannot escape your shadow. Effectively, to borrow from quantum physics, they would form complementary entities, neither one being fully describable without the other.

A metaphoric image of their differences may be more like the unity depicted in the Tao symbol by the complementary representations of

Yin and Yang. These are the familiar interlocking though seemingly separate pair of tailed swirls, which together form a complete whole, although each has a small representation of the other within them. For example, one could imagine our universe as having ordinary matter, and the shadow universe as having what to us would be shadow matter. Indeed, perhaps this shadow universe *is* where all the theoretical supersymmetric complementary sparticles of the new standard particle model exist! How could we know?

Superimposition

One might have imagined these vacuum-sourced, quantum-sponsored proto-universes as sitting right next to each other at the origin, like two peas in a pod. On the other hand, because of the explosive way these universes would have originated, one could also imagine them as being driven far apart. However, it seems more likely that they would be neither adjacent nor spatially separated since they started from one single quantum disgorging event, with no pre-existent spatial dimensions to move into. Therefore, with nowhere to go, and transparent to each other's physical expansion forces – until gravity kicks in later – they would remain co-existent at the disgorging point. They would be more than just co-located, though. Their superimposition would be a mutual, though segregated distribution *throughout* each other.

The symmetry-breaking mechanism, such as the development of alternate dimension sets, would likely prevent most physical interaction between these budding universes, except perhaps the interdimensional sharing of gravitational effects we discussed. Such mutual gravitational attraction between them would also tend to ensure their continued proximity within each other as they evolved into their respective expansion phases. To portray this arrangement, imagine such a binary-universe system as a pair of expanding bubbles, where one is concentrically inside the other, roughly the same size, though not touching. These universes would remain superimposed within each other, expanding from a common centre of origin, although maintaining dimensional separation.

Should this speculative shadow universe have developed along these lines, we might find some circumstantial evidence of it within our universe. Frustratingly, such evidence for our shadow could be right under our noses, yet effectively infinitely apart – with each superimposed twin being interspersed throughout the other's existence. Though dimensionally segregated, if gravity-bound, they could remain in gravitational lockstep as an integral part of each other's existence even though their evolving outcomes might be different. One only has to consider the energy-intense tidal effects that occur on earth due to the proximity of the moon's gravity to appreciate how effective gravity can be as an energy transmitter between massive objects. If gravitational effects were to interact between these two universes, we might be able to infer the invisible presence of such a shadow universe by its gravitational influence on our own universe's more massive objects.

Any gravitational interaction of large clumps of accreted mass-energy present in the shadow universe would tend to amplify the gravity-bound clustering effects among such massive objects as galaxies in our universe. That could happen even if the physical structure and mass-energy ratios of the shadow universe were quite different from ours. That could lead to similar symptoms as those identified as "dark matter" in the currently favoured, though speculative explanation for the curious clustering behaviours of galaxies in our expanding spacetime. As already noted, these galaxies seem held together by additional, unidentifiable sources of gravity, not accounted for by measurable mass and momentum of the star systems observed within those galaxies. Through the exchange of gravity-bound tidal-like forces, mass distribution in the superimposed shadow universe might gradually take on approximate correspondence with many of the mass concentrations in our universe, and vice versa. That would generally amplify the local effect of observed galactic-scale gravitational accretion.

Furthermore, suppose the expansion rates of these two superimposed, tidally linked universe "bubbles" were marginally out of synchronization with each other. If the shadow universe were to expand slightly faster than ours, throughout the last several billion years, it might induce an expansive, gravitational tug on our spacetime. Such an expansion-related tension among the binary world's galactic masses would forcefully drag our universe's expansion faster outward along with it. We would observe that effect as an inexplicable acceleration of our universe's expansion, which we might then identify as a mysterious form of dark energy. However, the two superimposed universes might just be oscillating in slow motion inside each other's fluctuating concentric, dimensionally transparent "bubble," each gradually taking its turn to be the slightly larger every few billion years.

Sufficiency of Reality

A different speculative model for existence contemplates a mix of realized and unrealized physical phenomena, which would contribute to the apparent form of our forever changing spacetime. A relatively small amount of manifest matter would exist with a large balance of potential mass-energy held unmanifest as unrealized pre-particles within uncollapsed wavefunctions. At one time, all the components of matter would have been unrealized, though over time, more of their wavefunctions collapsed, becoming manifest. This relatively small amount of tangible reality would represent the matter and energy that we sentient beings consciously interact with and observe. The remaining unrealized pre-particles remain as wavefunctions with no material form. They represent endless possibilities in a state of superposition, though nothing substantial.

An artist can use a broad brush on canvas to give the vague illusion possibly of clouds. It may suggest the potential for clouds, though only if the artist were to devote detailed creative effort to the edges, colour, and values within those potential clouds, could they become a realistic version of clouds. Without the artist intentionally providing that cloud-like detail, the broad-brush marks might instead give the impression of distant forests, mountain ranges, smoke, or an impending night sky, depending on the interpretation of the observer. Only the conscious awareness and intent of the attending artist and the imaginative observer finally render that potential into a representation of the reality of their imagined worlds. Before such focus occurs, in the eyes of the casual observer, they are merely suggestive marks on a canvas. Similarly, unmanifest possibilities of unrealized pre-particle potential may provide some suggestions of reality for casual observation. However, they would be of no directly determinable substance, although, even though unmanifest, their presence might still interact mildly with our manifest reality.

In ancient days, before telescopes and other instruments that extend the senses of our intelligent awareness and curiosity, when humans looked up at a night sky, what they saw was their immediate reality – an infinite number of tiny points of light that twinkled. The depth of their understanding of what they saw defined their realization of the night sky. That sky was the broad-brush illusion of reality that served as the limiting boundaries for the mindset of that era. No more was required of it; no more was known of it; it was sufficient.

As those sentient observers became more curious and more informed as to the likely nature of that twinkling sky, that ancient sufficiency of reality became inadequate, evolving into a greater depth of definition of detailed understanding. This enquiring nature developed to the present day, when cosmologists and physicists examine individual stars, attempt to define the known structure of the universe, and intensely contemplate the ways it could have come into existence. As sentient beings define a workable description, at the maximum level of resolution they can comprehend, then that becomes the new sufficiency of reality until a future understanding confirms or corrects it, creating a yet more detailed experience of form.

Naturalness

In the business of contemplating our origins, many speculative ideas combined with limited though expanding factual data are becoming available to enable better tentative modelling of our existential origins. While these ideas are stimulating, and many of them ardently researched, not satisfactorily addressed by any of them are those curious improbabilities associated with the values of the crucial constants of nature in our universe. These constants include cardinal ratios and parameters essential to the fundamentals of our universe's existence and dominate the way our physics works for us. Critically, those bases for the physics of matter that we have come to know could not vary by a significant amount without the whole edifice of atomic and galactic structure collapsing, and our very existence ceasing.

For a litany of technical reasons not detailed here, as far as we can determine, the existence of our universe appears to be a highly non-generic and improbable version of all possibilities that we could reasonably project. This conundrum leads to the "naturalness" question – whether our universe is natural or unnatural. This question is creating much heated scientific debate. At its core, the choice appears to be that if the universe is natural, it is inevitable, and our scientists should be able to derive an adequate description that models how it naturally came about – a "Theory of Everything" – which they have not yet achieved. On the other hand, if it is unnatural, it might exist for one of several reasons:

- A god's creation (the religious view)
- The consequence of a highly improbable set of circumstances spontaneously occurring (a random event)
- A unique outcome from an otherwise life-desolate set of universes (the multiverse scenario)
- A non-deistic agency encouraged it

I suppose it is conceivable that our universe is but one among countless random other possibilities in some giant multiverse of which only a few, including ours, could randomly develop the right parameters to support life and curious sentience. Nevertheless, science is reluctantly considering that there is such an extremely low probability of an outcome occurring that matches our existence by random chance alone, that maybe something else is going on – something more self-serving.

Scientists are generally skeptical of mysterious solutions, as well they should be. However, once you eliminate falsifiable hypotheses, what remains are the spookier possibilities that Einstein detested so. There may be more than the uncaring hand of random probability at work in the fine-tuned selection of these constants of nature. Our Goldilocks universe may be exactly right for sentient life because it must be. Possibly – and this is the scientifically unpalatable part – among the seemingly random, probability-inspired quantum events that appear to define the origins and structure of our physical universe and continue to do so, there might be room for conscious intent.

A tentative idea we touched on earlier relates to the possibility of consciousness influencing the outcome of a collapsing quantum superposition state. At the very edge of our scientific perception of physical reality, there appears to be an interface with not just ever-smaller particles, but with what seems a non-physical basis for existence. At the very least, it is one with incomprehensibly different physics from our experience. Such an unconventional concept might require exotic forms of matter to interface with novel types of fields, and currently at least, that would be in the realm of the metaphysical.

Among these possibilities could be a holistic network of quantum states entangled with a universally distributed field of unconditioned consciousness associated with the nonconscious processes of sentient beings. Such a dynamic system would likely permeate all spacetime, influencing form at the quantum superposition level, participating in, and promoting the physical state in which we exist. Our reality would then be floating in, and indirectly interacting with such an extradimensional field, which may be subject to influence by its resident sentient beings such as us and would sponsor our highly improbable spacetime physics to support the system.

If this were the case, what we think of as scientific exploration, what we imagine to be a discovery about the intensely detailed physics of an already-existent reality might instead be the cause of such detail that did not pre-exist its finding. Exploration, that spirit of enquiry, would thus function as creation because by applying conscious intent to exploring and understanding further and deeper, we unwittingly create the necessary detail to support that newfound, newly emergent reality. By employing coherent attention and conscious awareness to comprehend how we think the universe could have become, we resident sentient beings may cause it to be. And thus, perception becomes creation.

Next, we will examine the implications of this possible linkage of consciousness to reality and develop a directional model of how and why the universe evolved the way it did and continues to evolve still.

PART 3

Sentience

"It is not outer awareness,
It is not inner awareness,
Nor is it suspension of awareness.
It is not knowing.
It is not unknowing.
Nor is it knowingness itself.
It can neither be seen nor understood.
It cannot be given boundaries.
It is ineffable and beyond thought.
It is indefinable.
It is known only through becoming it."

Mandukya - The Upanishads C. 900-600 BCE.

PRIMORDIAL SENTIENCE

Can we imagine conscious intent on a universal scale? Here we will examine the association and role of consciousness as it concerns the foundation of perceived reality. We noted in Part 1 that the term "Sentience" describes a state of elementary or undifferentiated consciousness – a naive unconditioned awareness. It is also the propensity to perceive information from the senses, including, and of particular relevance here, the sense of thought. Primordial sentience could then describe a fundamental unconditioned consciousness evolving sensory perception.

Earlier, we remarked on the curious similarities in some behavioural characteristics and intimated relationships between consciousness and the quantum mechanical states of superposition and nonlocal entanglement. The view expressed here is that consciousness may take the form of a universally available resource that catalyzes the local processing functions of an individual mind. In such a holistic view, three distinguishable levels of sentient consciousness operate and interact with our reality.

- The most obvious level is local to each of us – our individual conscious and unconscious mind supported by our senses.

- The least visible, although the highest level would be a universally distributed sentience operating beyond personal awareness, behaving as a ubiquitous conscious field throughout spacetime – and perhaps beyond – and possibly influencing fundamental physical laws of existence at the quantum level.

- The intermediate level version of sentient consciousness takes the inconspicuous form of multiple individuated subsets of the universal kind; each one operating as an interface within the veiled part of every sentient being's nonconscious mental processes – their psyches. Mostly hidden from the local consciousness, it would act as a dynamic link between the nonconscious of every individual sentient being and the universally distributed version of sentience.

Local Consciousness

As used here, the expression "local consciousness" means the consciously self-aware mind within each of us, which activity appears to take place primarily, though not exclusively in the neocortex portion of our three-brain system. It is the illuminator of our conscious mind, and a receiver and processor of unconsciously collected and selected sensory information, including thought. Though not unique to humans, arguably it seems more developed in us than in other earthbound sentient species. Within the physical infrastructure of the sentient brain, the local consciousness primarily extracts meaning from information or creates information from meaning. When we talk of meaning, we embrace any mental stimulation to the perceived status quo initiated by local changes stemming from physical sensor inputs to the brain, self-generated ones, intuitions, emotional surges and spontaneously introduced stimuli originating from other sources.

Information is a universal concept, essentially requiring a change or variability in a perceived state. A perspective on something that does not change becomes uninformative. Something that frequently changes state, particularly if it provides a recognizable pattern, cause, or discernible progress in the process, can be highly informative. In this sense, information, like energy, shares many reciprocal characteristics of entropy. Entropy is the destination path of highly ordered states of information of minimum entropy as they degenerate toward a uniform low-energy, disordered state of minimal information and maximum entropy. Indeed, much of the math of information theory is common to entropy theory. Some information theorists seek to portray the entire universe as composed entirely of projected or extracted information – quantum information to be more specific – and they may not be wrong.

Arguably, from a practical perspective, information must be informative. Data only has real informational value if it enables meaning and thence prediction whereby an intelligent sentient being can observe, recognize, and respond to it. Information must be received cognitively, otherwise what might have been perfectly valid data, if unrecognized, has no better informative value than random noise. Thus, the limiting degree of information validity lies in both the technical generation of informative data and the competent reception of it. That role of competent reception falls directly or indirectly to consciousness and, perhaps to a lesser degree, the unconscious. The intelligent conscious processing of data into information and then into meaning enables predictions of, and thus possible influence on a sentient being's end-state with probabilities of success better than random chance, thereby increasing the host's evolutionary odds of survival. This advantage is in opposition to entropy, which continuously devalues information.

Once again, we see the ubiquitous role of conscious and nonconscious processes and interactions as actively fundamental to existence. These interactions not only cause insubstantial pre-particles in wavefunctions to manifest into particles of matter (by the act of observation), but they also create meaning from our surroundings by the analysis of information continuously extracted from packets of data. These roles of consciousness interrogating wavefunctions and data-packets to cause matter and meaning are remarkably similar processes and point toward the primal nature of consciousness in the creation of reality.

Our mind navigates us through life by projecting scenarios based on the intelligent apprehension of information extracted from the past and projecting potential future states of the probable and salient aspects of our existence. From the perspective of an individual mind, only meaningful information has any value; and then only to the degree to which it is capable of being comprehended and utilized. Our ability to survive and evolve, is dependent on the intelligent awareness with which our mind can focus on, extract, and interpret information describing manifested macro-reality, its potential interactions, and project its likely future state.

The scale of our sense of reality is dependent on intelligent information gathering and processing by local conscious and unconscious functions through our physical brain. How, then, does consciousness operate at such a high-resolution level within our grey matter? How can it have sufficient capacity to resolve the enormous cumulative amount of information portraying the history of our perceived reality? Neurological science tentatively speculates that within the brain's neural network, at a level below that of neurons and synapses, are large numbers of microscopic tubules, which act as resonant cavities, allowing a form of quantum computing to take place. Such a resource would presumably have much more information processing capacity than that implied simply by the capacity of the brain's neuron network alone. The concept is that at this level, quantum effects would allow for the vast number of awareness states necessary for the unconscious to process, interpret, and project our reality. This sort of research is exciting. It will undoubtedly lead toward a better understanding of the brain and mental functions, and perhaps the role of local consciousness within, though it may never explain the mind's sense of thought.

We recognize that understanding how to tune a radio does not mean we know how the orchestra plays the music or the intention behind composing the score. An ingenuous belief of promissory scientific materialism maintains that consciousness is an emergent property of the physical brain, which naivety equates to the expectation of finding the orchestra and composer tucked inside that radio. This faith vastly underestimates the actual form and role of the mind. An emerging realization turns out to be the other way around. The evolved physical brain may have been stimulated to develop thus, by the needs of the mindfulness of primordial sentience, the metaphoric equivalent of radios having developed to broadcast a composer's work as performed by an orchestra. The realization that sentience and thus consciousness is fundamental, and physicality is emergent from it, will undoubtedly turn scientific physicalism on its head.

Nonlocal Nonconscious

Nonlocal nonconscious is my clumsy expression for an extended part of the mind's unconscious processes. It is mostly hidden from the conscious mind, being part of the unconscious, and its source is not totally from within the physical boundaries of the brain – it is nonlocal to the brain. That qualification needs some elaboration.

The mind comprises both conscious and unconscious realms. A large portion of our mental process is generally hidden from our typical day-to-day conscious thoughts. The unconscious expends much effort analyzing data from our various senses and deciding what to do with it. Relevant portions of this data are passed on to the consciousness for further processing. The unconscious is also responsible for the full range of bodily autonomic administration and reflexes. It enables complex body-management processes that are not under conscious control and require constantly adjusting metabolism rates, hormone levels, pulse rate, and a thousand-and-one other essential bodily functions. Although mostly a hidden part of the mind's facility, these processes are local to the brain and, perhaps to a lesser degree, other parts of the body.

Also, in this class of hidden-but-local parts of the mind are the automated physical responses to everyday routine. These include learned bodily functions such as the repetitive muscle actions involved in running downstairs, which may not be so much hidden from consciousness as habituated out of its immediate awareness. Some of us may be familiar with the circumstance of suddenly thinking, partway down the stairs, about what one should be doing with which foot. One's conscious mind cannot keep up with the required speed of correct muscular responses, usually assigned to the unconscious mind. The resulting painful failure to navigate the steps provides evidence of our inability to perform consciously what has been a simple unconscious activity since infancy.

In contrast, nonlocal nonconscious processes are part of the hidden mind, often associated with the concept of psyche, spirit, or soul, which can recover information not available locally through conventional means. It provides the mind's access to emotion-borne faculties such as instinct, intuition, morality, and conscience; inspirational and spiritual experiences, including awakening and *nirvana*; and possibly any genuine "psychic" phenomena that fall under the catch-all umbrella of extra-sensory perception. This nonlocal part of the nonconscious is likely to operate beyond the conventional quantum-mechanical level of physicality and information processes and is nonlocal in the technical sense of being unrestricted by spacetime limitations. Available to it as a resource is the historical experience of the individual mind, the collective sum of all sentient experiential histories, and the creative potential that inspires the universe.

Nonlocal nonconsciousness can be likened to a hologram, whereby laser light beams illuminate an object from specific angles, and the resulting interference pattern from the scattered light field is recorded. When that recorded image is illuminated and viewed from similar angles to those of the original laser beams, the object's image emerges as a three-dimensional, stereoscopic hologram, which may appear to move as the viewing angle changes. Sold in novelty stores as anything from buttons to bookmarks, sophisticated versions of holograms appear as security features on credit cards and banknotes.

Importantly, recording interference patterns rather than the image enables the whole picture to be present in every part of the hologram. Cut off a piece of an authentic hologram, and you can see a fainter scaled down though complete holographic image in the cut-off section and the remaining part shows the same whole image. We can apply this analogy to the individual nonlocal part of the nonconscious. The individual psyche or spirit represents a fragment of the larger "original" holographic image that corresponds to the universally distributed, shared consciousness concept. The whole hologram is present in every individuated fragment; only the scale and intensity are different. In a dynamic sense, every experience of one part reflects on the whole, and vice versa. When we talk of spirit, psyche, or the soul, it is this hidden nonlocal part of the nonconscious mind to which we intuitively refer.

The idea of such a nonlocal, dynamic life force is not new. For example, it loosely corresponds to the ancient concept of Ch'i or Qi which has been around more than three millennia, and refers to a life force associated with, but able to exist outside the body. Of similar age and meaning, originating out of India is the Hindu notion of

Prana – a life force or life energy. Related concepts have arisen in other ancient cultures. It seems that we, as a species, may have intuited a lot more about this subject in the past, than we allow ourselves to think we know now. All too often, our materialist-infused conscious mind blocks or denies uncomfortable, apparently counter-factual, and perhaps inconvenient insights. However, when in meditation, we sometimes manage to draw back the obscuring intervening veil by dampening the diverting noise of local conscious thought. Then, it is through the access provided by the state of nonlocal nonconscious that we sentient individuals can occasionally glimpse the potential of all that there is – and there it is: universal sentience.

Universal Sentience

Entropy is the tendency of hot things to cool, batteries to fade, fabric to wear out, eggs to break, and living things to die. In technical terms, entropy is a sequence of energy states transitioning from higher order toward lower order and is linked to time's arrow. However, included within those seemingly inescapable entropic changes degenerating toward universal exhaustion, in the chinks and gaps between those inevitably devolving shifts lies a myriad of unlikely opportunities for the opposite to occur. Such opportunities await the evolution of complex systems to reverse the march of entropy – albeit temporarily and locally – and instead, increase local order while deriving energy needs by diverting or accelerating other entropic processes.

Thus, within the inevitable entropic backdrop are found opportunities of available, useful potential energy, allowing complex systems to develop, self-organize, and to create pockets of higher order and evolve life itself – a process termed autopoiesis. Such pockets of creativity stand proud from the general entropic decline. They metaphorically push water uphill, creating striking contrasts of potential amid an overwhelming, enervating descent of the universe into total disorder, heat exhaustion and oblivion – the entropic death spiral.

Ironically, though, if it were not for that massive energy sink that is universal entropy, there would be no such instances of these potential energy changes available to exploit. Without them, there would be no energy gradient to provide the opportunity for moulding the self-organizing pre-cursors of life. These oases of potential in the vast space of the universe are where the imbalance between the overwhelming laws of entropy, and the tendency of complexity to evolve into sentient life, against all odds, favour complexity and sentience. An indefinite cycle of sentience-benefitting probabilistic physics conditions the energy from quantum fluctuations, thus inducing atomic matter from energy, molecules from atoms, living cells from molecules, and sentient beings from living cells – consciousness manifested.

Imagine a boundless universal field of unconditioned conscious energy saturating the spacetime-dimensioned medium of the universe. It confers local sentient perception with potential scenic backdrops of reality and is a stage for existence's dramatic performance. The nonlocal nonconscious of each participating sentient being represents the stagehands and the actors, filling in the metaphoric scene with props, costume, and action. Through interaction with this field, the totality of individual sentient experience and comprehension indirectly teases reality from potential, manifested out of emptiness. Each sentient being's nonlocal nonconscious represents a dynamic representation related to and reflecting the entire universally distributed field of undifferentiated consciousness in which the conception of existence resides.

Any continually creative process or complex system requires performance feedback and informed change, and that is what we have in the continually changing tumultuous bulk of the universe. Only by requiring changes to happen, sequenced through the illusion of time, can existence form, consolidate, recreate, and evolve. Only through unrelenting change, can the collaboration of a universally distributed consciousness and the myriad of individual feedback components – all sentient beings – transform conceptual potential into local manifestations, and in so doing, grant subjective reality to those sentient beings inhabiting them.

Within this description, the individual nonlocal nonconscious provides a point of focus, an informational bridge with the universal-scale consciousness. Thus, part of every individual's nonconscious state connects to and reflects the more global state, that field of universally distributed consciousness which humanity has labelled in many diverse ways over the ages. Such labels encompass ancient spiritual expressions, such as Great Spirit and Holy Spirit, and more secular terms such as Brahman, Cosmic consciousness, and the collective unconscious.

The expression "universal sentience" is no more correct a label than those are, but it better describes my intuited, subjective image of that source of a universal driving force. In this choice of expression, the adjective "universal" embodies the sense of a field of presence at least as big as the universe – although dimensions of size will not bound it – and "sentience" embodies aspects beyond just information processing, such as an evolving conscious awareness and intelligence. However, since adjectives tend to add bulk, the term "Sentience" will generally be used herein, with the upper-case S denoting its universally distributed, nonlocal quality and stature.

Conceptually then, the nature of Sentience is as an extradimensional field-like presence. It represents a ubiquitous universal creative force that shapes form by influencing probabilistic quantum state potential. Overall, Sentience acts as a fundamental ground of being, within which our entire physical and metaphysical realities arise. More than a larger version of an individual's sentience, it is an all-encompassing field of distributed unconditioned conscious awareness. It forms the common source from which those individual sentient psyches and physically manifested states spring forth, and which are accessed through the nonlocal part of each unconscious mind.

A sentient being's mind, enabled through the individual's nonconscious access to Sentience, can unconsciously acquire veiled awareness from Sentience and translate it into intuition-driven instinctive action or forms of intuitive insight for processing within consciousness as thought and inspiration. Once engaged by the individual, expressions for this nonlocal state of nonconscious may be characterized by such psychic qualities as intuition, insight, instinct, extra-sensory perception, awareness, and enlightenment, and may also include the many unremarkable, everyday perspectives on our tangible reality.

The vital link between Sentience and an individual sentient mind provided by that unique interface, the nonlocal nonconscious, is the focal point representing psyche or spirit. That is why cultivating a voluntary means of establishing a quasi-conscious dialogue with the unified ground-of-all-being through meditation is so significant. Once awareness and access to the universal comprehension of Sentience has been established on an individual basis, the scales of ignorance fall from our metaphoric eyes and an understanding of who we are and what life is can flash into our conscious mind. Providing we do not reject or corrupt that understanding, nothing is ever the same again, and while we can always go deeper into more profound states of awareness, we can never return to our prior state of afflictive ignorance.

Sentience Obscured

Sentience can be envisioned metaphorically as a diamond jewel, with each of us being a facet among many; each one consciously facing out though nonconsciously reflecting what is within; each one a key part of the exterior, illuminating but dependent on the interior – and indeed on all the other facets. The facets are an integral and essential part of a diamond jewel; without them, it is just a crystalline lump of carbon. The jewel can only be fully described through reference to the sum of all its facets. Each facet, the mind of a sentient being like us, is a direct path of communication, a bridge to the gleaming gloriousness of the total jewel. The reflections of Sentience stream out to us, to everything, everywhere. To the extent that each is able and allows, sentient beings can sample from this pure resource of potential. Sentience – omnipotent, omniscient, and omnipresent – is always available to us, not because it is there and we are here, but because we are within it, as fish in the ocean. We are enfolded, enveloped, entrained, and integrated within this metaphysical intelligence. We *are* it. Moreover, it is not an it, or a who; it has no form, it just is.

When each attenuated facet of Sentience interacts through individual psyches, those minds veil most of that exposure from their consciousness. That veiled condition of ignorance protects us from the disorientation that might result from continuous awareness of Sentience, constant access to its omniscience, and ceaseless cognizance of the illusion of the virtual reality of our daily lives. Indeed, we would probably be unable to fulfill our evolutionary purpose, or ourselves, if we were unremittingly in a state of full comprehension of why things are the way they are, rather than how they seem to be. Undiluted access to Sentience would overload the individual sentient conscious mind to total distraction; we are just not ready for that. That is why our real metaphysical side – our spiritual nature, our very core of being – remains mostly obscured from our day-to-day awareness. For each of us, its point of access to the individual mind remains veiled in the shadow of our unconscious. In certain circumstances, however, this veil may lift, offering a glimpse of the wealth of Sentience.

Much of the fabric of this veil, which obscures us from overwhelming exposure, but which also inhibits our understanding of who we are, consists of mental noise – mindless thoughts and egocentric preoccupations. The term noise applies here in both its literal and technical meanings. In a literal sense of sound, there are constant external distractions such as voices, music, traffic, and other ambient sounds that we welcome, endure, or ignore throughout our day. The more technical meaning of noise refers to anything that devalues or masks information that might otherwise be available for processing. It can be present in any form of medium – acoustic, visual, electromagnetic, and mental awareness – and includes all kinds of dissonance, interference, or incomprehensibility. This broader definition of noise represents strands in the fabric of the veil, interwoven with ignorance, diversions, and – the main offender in terms of meditation – the endless chatter of one's egocentric conscious thoughts, the monkey mind. Only when this most significant source of noise quietens can there be a possibility of catching a glimpse past the veil and witnessing even a small part of the grandeur that is our shared Sentience.

This psychological veil may serve to protect us at our current stage of maturity; however, it also inhibits our ability to conceive, comprehend, and feel as one with our common source of being. Ignorance of our ground of being can induce a sense of isolation, and even if having some awareness of Sentience, it can still promote a feeling of separation. This ignorance of both origin and purpose can contribute much to our societal and individual psychological pathologies. The sense of separateness from society and its members leads to disassociation, abuse of all kinds, and the development of aggressive and discriminatory philosophies. It fosters a sense of us as being set apart from nature, life, environment, even each other, and of each of us being an exclusive entity, severed from the rest of the universe through which we travel – none of which is true. Fortunately, the impermeability of this barrier is not irrevocable in this respect. The disconnect only remains if we continue embracing ignorance and fail to awaken to a glimpse beyond the veil.

Meditative insights, those glimpses beyond the veil, reveal our purpose — and that of every sentient being — to be chiefly two-fold. First, it is to be an involved and evolving sentient manifestation within the whole universe creation — an actor on the stage, a cameo within the work of art. Second, and more importantly, it is to function as a two-way interface between our hands-on experience of physical existence — our reality — and the metaphysical Sentience. Primarily, we interact consciously with existence and nonconsciously with Sentience as a sensor capable of recording experience, providing situational feedback, and indirectly helping influence outcome — an agent, if you will.

The primary role of all sentient beings then, as interface or agent, is to recognize the need for and complete this experiential feedback, to act as observers and enablers — as sensors — in a co-creative role with Sentience. This role is nothing less than an executive agency in the joint micro-completion of the macro manifestation of the universe, conjured within the field of Sentience. We — all sentient beings — are the hand, the eye, and the work of this ultimate creative artist, which is also us. Other than that psychological veil we spoke of, there is no separation between our nonconscious minds and Sentience; we are as one. We are at once the artist and the art, the dancer and the dance — the creator and the created.

Of Self and Spirit

We noted earlier that the notion of self consists of the physical body, the conscious and unconscious mind, and that hidden part of the nonconscious mind represented by psyche or spirit. These essential components, colloquially body, mind, and spirit, make up the individual self. The individual's physical and genetic makeup, learnings, life's experiences and chosen responses to them, are the main contributors to the unique, though impermanent characteristics of that individual. It is a misconception that because the combinations of these facets making up the unique individual's identity seem to be locally associated with that individual, then the self must be entirely contained within the body, separated from others and the environment. However, separateness and uniqueness are not synonymous. Our body may seem independent from others because it is free to move around, rather than being stuck in one place like a tree. However, it is still clearly dependent, both physically and psychologically, upon others and communally shared material resources, such as atoms, air, water, heat, food, the environment, the world, and the universe.

A global perspective discussed earlier was that the individual psyche or spirit could be considered a segment or facet of universal Sentience. Everyone comprises a body, mind, and spirit, with spirit being a shared quality between Sentience and the individual's nonconscious mind, unconstrained by the dimensions of spacetime as in the principle of nonlocality in quantum mechanics. Nonlocality does not imply that the psyche or spirit is remote from the mind, but that, though associated with the individual, some interactions are not restricted by local physical spacetime limitations. Spirit is the agent, the holographic image reflected at the access to universal Sentience, the individual's unseen, and impermanent taproot into the ground of all being. Similarly, all other sentient-selves form segments of Sentience; individual facets of all-that-there-is, embedded within and forming an integral part. Although there may be physical space between mobile bodies, there is never any separation between individuals, their environment, or their nonconscious minds because they continuously link with their shared universal environment and their common ground of being.

These notions of self are attempts to describe the concept of an individual resident spirit in pragmatic, secular terms that are neither arcane nor mystical. We noted earlier that the terms psyche, spirit, and soul are just labels applied to similar meanings, albeit with some opinion-nuanced differences. Anyone uncomfortable with the term spirit or soul can substitute that of psyche and will get the same basic understanding. Psyche is a term adopted by the psychological sciences. Although sounding the more secular term, all three refer to the inferred hidden driving force of a being – the actualization of the individual. The notion considered here envisages individual focus points of psyche or spirit, within a universal field of *uber*-consciousness. Think of each psyche as a portal or node, one of innumerable nonlocal elements associated with sentient minds and generated within the ubiquitous field of universal Sentience. Each of these elements, though nonlocal and certainly non-physical, temporarily entangles with part of the nonconscious mind of each sentient being for their lifetime.

As we also noted, in some religious interpretations, an individual soul-entity is deemed to post-exist an individual, and at the death of the body, the soul purportedly exits intact into a heavenly realm while retaining some recognizable form and character reminiscent of the now-dead individual. The contrasting interpretation favoured in these pages is that the terms psyche, spirit, and soul all conceptually denote the same entity, which is locally impermanent. It is a nonlocal projected representation of Sentience, and it ceases its interaction with the individual's mind and body at their death. Once the body is deceased, the spirit disengages from the disintegrated character of the now-dead individual. As an internal fragment of universal Sentience, its association with the individual sentient being concludes, and it returns to Sentience – as a raindrop falls to the sea. The sum of all its locally gained experiential information unconditionally accumulates within a common repository – the ocean of Sentience.

Analogies of Spirit

That raindrop metaphor is a useful analogy to the concept of spirit; it is both simple and reflects a natural process. However, i want to mention another metaphor, which, at an early age, became for me a philosophical insight into the nature of our existence. It was the childhood experience described in the first chapter, which still resonates with me today, and, in retrospect, provided the memorable experience of recognizing oneself as being part of a metaphor. Briefly recapping, the experience was of a Camera Obscura, a pinhole camera the size of a room, and i was inside it. The magic of it was that there was a dynamic live image of the outside world, projected pale upon a screen inside the darkened room, with no apparent means of projection. My keen child's eye eventually found the source of the image, a piercingly bright pinhole in the sidewall of the darkened room. To an eye accustomed to the dim light inside the room, the pinhole was dazzling although it seemed impossibly small. However, i knew – because i had gone out the door to check – that if i could squeeze myself through that pinhole, i would engage with a dazzling reality, the source behind the pale projection i was experiencing.

As a child, without knowing what a metaphor was, i intuitively felt this to be a lesson in philosophy, had i known what that was! I recognized the representation of two realities, the pale one we live in, and the one we were mostly unaware of – the brilliant one beyond yet projecting this world; the one that was complete and real, the source of our pale existence. I found out much, much later, that similar philosophical observations were recorded millennia ago, dating back at least to Plato – so, not a novel metaphor! I have since come to think of this as a simple metaphysical analogy describing the relationship between an individual's mental experience of reality and universal Sentience. In this interpretation, Sentience is the dazzling All, and each sentient being's corporeal psyche, although graced with local experience, is a pale though authentic projection of it.

Interestingly, recent scientific musings, based on black holes, holography, and information theory, propose that our reality could indeed be projected information. From the study of black hole entropy, it turns out that for any given volume in spacetime, the maximum amount of information available to describe the contents therein is proportional to what could project onto its outer surface area. Initially developed from studying black holes, the theory can apply to everything from atoms to universes and has been extended to speculate that our entire spacetime reality might be a holographic-like projection of information located beyond the boundary of the universe. The variation on this view as proposed here is that existence is not a pre-existing projection permanently available for conscious apprehension, but a dynamic extraction continually selected by Sentience from an inventory of all possibilities held in an unmanifest state of superposition.

These kinds of analogies illustrate that, despite each of us being unique, no sentient beings are individually isolated bodies, minds, or spiritual entities. Nor could a sentient being – body, mind, or spirit – exist outside of, or independently from that universal, unconditioned consciousness of Sentience. Sentience is all that there is, and we are an integral constituent. In attempting to attach a label to such an unknowable fundamental source entity, we derived sentient beings have tried throughout the ages to categorize it with a thousand-and-one reverential and awe-full names. Some labels seek to personalize, some objectify, and some distance. Nevertheless, they are just labels, and inadequate ones at that. Categorizing anything with a label tends to distance us psychologically from the entity being labelled. Such distancing limits our realization of being a participant in that which is

All. For, what we are calling universal Sentience here is nothing less than unbounded *all being*, containing the thought of our universe and all within it. It is the eternal creative source from which, and in which all existence springs. How could you realistically categorize that?

Manifestation of Form

Although Sentience cannot be categorized, we still need to attach some label to the concept so that we can economically think or talk about it. Despite the dangers of labelling, an appropriate label should perhaps avoid reinforcing ancient superstitions or fostering the egocentric notion of it being just a resource that our egos may use to assure security, longevity, or self-aggrandizement.

In the interpretation offered here, Sentience is centre stage and not just centre stage; the whole stage, theatre, everything – and every non-thing – and we are part of it. We conscious creatures, along with our entire physical, mental, and spiritual existence, are part of and interact with Sentience. Together, we create reality wherever we focus the spotlight of awareness and intent. Our on-stage "reality" manifests as it does within Sentience, though we are a contributory part of this illusion – this virtual universe. To Sentience, of which we are a part, we are sensory perception. Our reality is part of its lucid dream, so, to us, who measure it by reference to the same modality, it seems solid, real, and full of physical form.

Imagine if the characters in a video game were to become self-aware. They would consider the other patterns on the screen as their "reality" because they consist of the same raw stuff as themselves. Anything they could conceivably develop to detect, measure, or manipulate their "reality" would also be made of the same microscopic material and would merely serve to prove it real to them. Observing them from beyond their dimension-limited screen, we would know otherwise, since we are outside the dimensions of their "reality," seeing it as nothing more than illusion manifested by transitory exchanges of electrons and photons on a two-dimensional surface. Our three-dimensional "solid reality" includes the video game's two-dimensional version of their "virtual reality," so we can know what the video characters cannot. However, their "reality" does not include ours, which remains unknowable to them.

Most of us believe that ours is the only "solid reality." We would be quite sure of that and deny otherwise. Even so, driven by our curiosity and probing consciousnesses, scientific and philosophic doubts have arisen regarding even our physical reality. As we have seen, the more we examine our reality in detail, the less tangible it seems, until we must admit that, at the leading edge of our sciences and our most profound secular spiritual understandings, the underlying form of our physical reality is but smoke and mirrors. It is inherently empty. Upon close inspection of the fundamentals of our world, there is no intrinsic form, no permanence, and no essential substance.

Our most current scientific explanatory theories include a reality driven by microscopic extradimensional superstrings, quantum mechanical appearances of insubstantial particles within multiple fields, and holographic information projected or extracted – from where we know not. Doesn't that sound rather like those video game characters that may begin to speculate that their photon-based reality could depend on some unknowable source projecting from outside their world, intelligence beyond their screen-limited horizon of reality? The foundational microscopic components of our physical existence are beginning to seem as fleeting as theirs. Our house of reality is built not even upon sand but on lucid dreams.

Consider then, our material reality, its form, its limitations, and its impermanence. How might we be influencing Sentience in co-creating this perceived existence? Could that profusion of largely unrecognized, sentient-focused nodes of spirit populating this virtual universe be collaborating in shaping the fundamental physical manifestation and evolution of our reality? Let us see how such minds might influence matter.

MIND OVER MATTER

Influencing Quantum Phenomena

The role of sentient beings emerges as being a creative tool of the universal force of Sentience, not as mere experiencers of a creation process, but as executive agents. One might question how it might be possible for us small sentient beings – at least relative to the star system in which we exist – to influence something the size of the universe. Conventionally we talk of microscopic or micro scale as meaning the exceedingly small, and macro scale meaning the large, but we less often refer to the mid-scale range. However, as measured by orders of magnitude of mass, we humans do lie right at the centre of a universal scale range.

In our immediate vicinity, the largest single mass, say an average-sized star like our sun, weighs in the order of 10^{30} kilograms. One of the smallest typical individual masses, such as an atom of carbon, weighs in at some 10^{-26} kilograms. Humans typically weigh in the 10^2 kilograms range, which, mathematically, is roughly midway between the masses of atoms and stars. Furthermore, through the creative development of our exquisite physical forms of remote observation and influence, we can extend our reach of awareness substantially in both directions, down into atoms and up to stars. So perhaps somewhere in that mid-scale range might indeed be just the right size for a smart tool engaged in co-creating the universe.

We noted that the thought-full veil, which obscures an individual consciousness from its nonlocal nonconscious psyche – and thus from conscious insight into universal Sentience – is permeable in one direction, though in the other, not so much. In a manner somewhat like the transpiring waterproof clothing that lets water vapour out yet prevents water from coming in, the veil only works as a barrier in one direction. Our consciousness is transparent to Sentience via our unconscious; all our personal experiences, haphazard emotional states, irrational, and rational conscious thoughts, and all our sensory experiential feedback are available to Sentience. However, our consciousness is not usually privy to either universal Sentience or even most of our unconscious. Except, that is, for those occasions of awakening when the veil does become somewhat permeable in that direction also.

Improving the likelihood of some conscious access to Sentience requires us to meditate, to quieten the noise of our conscious thoughts, to cut through mental distractions, to surrender to the moment, and then to manage the memory of any episode that may occur. The term "manage" is perhaps misleading, in that it implies an element of control – not the case here – whereas the intended meaning is more in the sense of coping and comprehension. Moreover, as observed earlier, any deliberate attempt to "manage" the fragile experience of awareness would not be successful if exercised concurrently, since that conscious mental activity would collapse the access to the revelations. Coping and comprehension can only take place after the session, based on recollections of the event. A successful episode of insight into Sentience would imply that one must have been able to relax, be open, quieten discursive, egocentric, or random thoughts, prevent distractions, and focus on manageable segments of the experience. Later, one can invoke good memory recall while using one's perspectival and conceptual abilities to contemplate the experience.

Another way that limited access to Sentience can be incurred is unintentionally and initially, nonconsciously. As noted earlier, this is likely the basis of instinct, gut feeling, intuition, inspiration, conscience, lucid dreams, miraculous revelations, and perhaps even the entire suite of authentic extra-sensory perception effects. That is when the relevant information and the impulse to act or not transfers from the nonconscious to the conscious mind, or sometimes even directly to the body's autonomic reflexes in command form without conscious awareness of the delivery path. These forms of accessing Sentience through one's psyche, although not particularly predictable, are part of a two-way path. The return path – continuous information flow from the individual mind through its unconscious and nonlocal nonconscious to Sentience – is mainly experiential feedback but can also include intent.

The formation and communication of intent via this route can be subverted by impure motives, ambivalence, or ambiguity. We may also pollute the potential purity of our nascent intentions with superficial, ego-centred, narcissistic, and materialistic distortions. There would be little coherence and depth to such expressions of intent, which may often take the form of muddled, ritualistic, ego-centred prayers or fickle desires.

Furthermore, many of us express our desires both to ourselves and externally to others, in equivocal or negative forms. Negative numbers are intellectual inventions of the conscious mind effective in math, science, and technology, but are not real. There is no such thing as a negative apple, for example. That only makes sense by imagining an apple and then attaching a "not" or negative logical qualifier to the image. However, consider this notion: the unconscious mind cannot comprehend negative logical qualifiers such as "not," "none," and negatives, finding them meaningless. When associated with a mental representation, the unconscious would grasp the representative image but not the negative qualifier. (For example, in a golf game, if a beginner is focusing on *not* going into the water hazard, the unconscious may ignore the *not* and instead focus on the image of going into the water, prompting their body actions accordingly! More skilled players, on the other hand, will ignore the water hazard, and focus positively on getting to the green.) Thus, a negative form of expression may produce a result in complete opposition to the original intent.

For the effective presentation of experiential feedback information and intent to Sentience, internally voiced expressions might best be authentic, coherent, intense, and pure. They could also benefit from being positive, consistent, enduring, reinforced through repetition and by many sentient minds echoing the same coherent information. Consider this to be like an ideal electoral voting process, where the confused and apathetic invalidate their voting right, and only the unambiguous votes sway the democratic day. A critical factor in the anecdotal effectiveness of authentic prayer and the outcome of minor miracles might arguably be coherent intent, though imagination and the placebo effect may also play a part.

Still, the question remains: even if we are clear and coherent in our intentions, how do we mid-scaled sentient beings of this world and any others, influence the entire evolving universe? How could we possibly be mentally affecting physics and co-creating our existence? Firstly, consider what might seem blatantly obvious – by converting mental intent into direct physical intervention. Secondly, and less obviously, is through the subtler mode of indirect mental influence.

Physical Intervention

For humankind, probably only within recent centuries would an awareness of anything beyond our immediate earthly territory (such as the rest of the universe) present itself for our minds to contemplate.

Only since pictorial representation and speech, probably less than forty thousand years ago, have discussion and exchange of thoughts and opinions about our existence and the environment even been possible. On the 14-billion-year timescale of the known universe and the four and a half billion years of planet Earth, this is but a flash.

Several centuries ago, most people thought the world was flat, fixed at the centre of a revolving sphere studded with pinpoint lights, and created in a week about six thousand years ago by a mystical old Caucasian man in a white robe. We also thought that the most massive object in existence was the Sun, which circled the Earth, and the smallest possible object was an indivisible solid mote called an atom. Now, in little more than a couple of centuries, we have learnt so much more about our physicality, and we still know blip on the scale of what remains to be known.

Yet, for better or worse, by being here intending, questioning, exploring, and creating structures, civilizations – and unfortunately, waste – launching our sensors and ourselves into inner and outer space, and consuming energy, we are irrevocably influencing the evolution of the planet on which we live. To a lesser degree, those influences now extend into our modest corner of the Milky Way galaxy, with some of our sensors now having left our solar system to explore the deeper interstellar space. Provided we do not prematurely terminate our existence, we will undoubtedly continue to influence and co-create our small corner of the universe physically. That may seem self-evident, but how have we been able to progress thus far so fast, some might say, so dangerously fast?

By way of our neocortex, we apply our sense of curiosity, discovery, and invention to everything that touches our awareness – attributes deriving from the creativeness of universal Sentience. Many if not most of the advancements that we have made are, at their root, inspirational responses to innate curiosity.

We might like to believe that the leaps in our scientific progress arise wholly from rational, linear thinking, and diligent effort. While it is true that such industrious effort is a necessary component providing the focused conditions to support a breakthrough and consolidate the resultant findings, the bulk of discoveries frequently develop from inspirational, intuitive origins, and often entirely speculative out-of-the-box mental leaps. These likely enter our conscious mind's realizations and conceptualizations as the fruits of nonconscious access to Sentience. As individuals or teams, much of what we contribute to the evolution of our universe originates from nonconscious access to the unrealized informational content of Sentience. Even when we abuse that discovered knowledge by labouring to build fusion bombs instead of generators, we are still inspired through Sentience.

Slowly, and with many backward stumbles, as a baby learns to walk, we progress our society and knowledge. Not uniformly either – some sentient members evolve faster, and some slower than the average; some are inspirational and some pathological. We can and do become aware of mistakes and even attempt to correct them to some extent, or at least not repeat them. Too often, we appear unsuccessful in that respect, but we can succeed. On balance, we sporadically evolve our society and our knowledge.

Conscious Intention

A subtler way of modulating reality is through mental means. Acknowledgement of this avenue is sparse, though at the leading edges of physics – specifically our comprehension of quantum-mechanical events – there is now discussion about possible influences by consciousness on the state of matter. Some reports have emerged on the apparent effects of mind over the behaviour of subatomic matter – minor, sporadic, and elusive impacts to be sure – though claiming some statistical credence and with huge ramifications. As noted earlier, a strange quantum-mechanical characteristic of subatomic pre-particles is that the state of a subatomic particle in superposition cannot be determined. It cannot even be considered to exist until its quantum wavefunction has been collapsed through interaction - such as in an act of observation.

The pre-particle in question does not appear to participate in our physical reality unless we permit it by becoming conscious of it. One could speculate that it remains unmanifest until drawn into a specific state of existence by the need to participate in an interaction witnessed by consciousness. That invitation could be through conscious intent influencing the innate and probably dynamic probabilities within quantum wavefunctions. We need to dwell on this latter phenomenon since it may be the key to appreciating how the universe might co-create with consciousness as an integral whole.

In early-recorded experiments, it was noted that if an apparatus was designed to detect particles, then a particle-like interaction was observed. However, if the experiment was intended to identify a waveform, then the outcome registered in the form of a wave. After some confusion, this led to an appreciation of one quantum quirk being that the intention behind the detection appeared to influence the likely form of the outcome. Other examples of similar evidence may be less robust though still claim some statistical validity, such as the apparent conscious mental influencing of the probability distribution of random numbers produced by an electronic random number generator.

That linkage is an aspect of quantum physics, one of several, which sits uncomfortably with most scientists. So little is known, and even less admitted, about the subject of consciousness or its possible influence on matter, that it is an embarrassment. As a result, among many staunch physicalists, it has become their challenge to pursue ever-complex solutions provided those do not include consciousness as fundamental. Such skepticism is not unreasonable since any such witnessed influence has generally been an inconsistent and subtle phenomenon. Incidents of such alleged consciousness-based quantum influence typically occur through somewhat circumstantial, statistically garnered, or anecdotal evidence. They are sporadic, essentially unrepeatable, and unfalsifiable in the traditional "scientific method" sense. Still, despite these difficulties, this avenue of research may ultimately prove productive and even necessary to further our understanding of existence.

Mechanism of Influence

We noted that the probable states of a pre-manifest quantum entity in superposition are described by Schrödinger's wavefunction equation, although the eventual end-state remains unknowable until observed. That formula expresses all the potential states that a possible pre-particle may be in before an observational interaction collapses the wavefunction. The equation describes a spectrum of probabilities for each possibility of finding the particle in a particular state. Hence, the wavefunction equation can be said to encompass the particle's realizing-potential – its possibilities. By squaring the function, it can yield the probability of each possibility occurring. Since the particle cannot exist in any material form until its wavefunction collapses, the wavefunction envelope *is* effectively the virtual pre-particle. Therefore, when we talk here of a pre-particle, we mean the wavefunction containing all the possibilities for the particle-to-be. Within the wavefunction, any-and-all possible end-states are in superposition, and nothing physically exists.

Consciously interrogating the wavefunction can trigger and influence its collapse. How sentient consciousness might influence quantum behaviour is unknown. Nevertheless, we can speculate on what might represent, directionally at least, an analogue for consciousness-mediated influence on pre-particle manifestation behaviour. Intuitively, the point at which consciousness, or more likely the nonconscious, might effectively influence particle manifestation is as it is about to emerge from its all-possibilities superposition state, while its outcome remains a probability-weighted possibility within an ocean of potential.

Although speculative, the notion discussed here is that the coherent projection of intention can influence the probability spectrum of outcomes for the possibility-defining wavefunction. This effect may be due to the entanglement of the pre-particle wavefunction with the nonconscious processes of the observer. We should, however, remind ourselves that the notional impact of an observer on manifesting reality through nonconscious intention pertains only to microscopic quantum effects. This depiction of a quantum-level, reality-influencing process does not suggest that conscious intent could produce immediate changes in macro-scale objects or events. It does not grant licence for magic or miracles.

Conscious Entanglement

The notion of the mind indirectly influencing quantum mechanical outcomes is philosophically difficult for contemporary science to take seriously. Even a hypothetical explanation is avoided because its inherently tenuous nature lies mainly in the metaphysical realm. As already noted, quantum mechanics is now used in nearly all aspects of modern life and is very well understood in a "black box" kind of way. That is to say, while the relationship between the input and the result is determinable, what happens inside the "black box" – the unseen wavefunction – to create the well-established, though probabilistically-ordained outcome is less well understood.

One model proposes that the quantum pre-particle in superposition within its wavefunction can trace all the possible ways for achieving every possible end-state at collapse. The numbers of ways of achieving an unlikely, complicated, or abnormal outcome are a lot fewer than the multiple paths leading to a highly likely outcome, which would naturally influence the probabilities of various possible end-states occurring. With entangled pre-particles, all those possible paths are shared or mutually correlated between them. It appears that each of the pre-particles mutually identifies the unique dynamic path to be taken at collapse, which leads to the correlated outcome of each of the pair. It is as though the entangled pre-particles were accessing the same playbook.

Seeking to clarify the conscious-observer role in this wavefunction-collapse mechanism, some have suggested that a computer could conduct a blind quantum state experiment remotely and automatically. Without the consciousness of a human observer being present anywhere in the loop to bias the result, they speculate that the twin state's outcome would turn out to be randomly unrelated, not correlated. The technical problem here is conducting such a blind experiment. Consciousness is involved in all aspects of planning and setting up the test, and then in observing the final report of the result in whatever manner it displays. It therefore becomes impossible, even with a computer randomizing the observations, to show that finally consciousness is excluded from the whole entanglement loop. All such experiments have failed to show any randomness of correlation between entangled pairs of particles. However, neither have they been able to prove the total exclusion of the subjective mind. The entangled particles consistently correlate with their complementary states, regardless of circumstances. Thus, a successful falsifiability test has yet to be devised.

The counterpoint to this outcome is that such experiments do provide circumstantial evidence that consciousness will always entangle with the setup. The entanglement will include any equipment used to determine the quantum state of the particles. The effect can even be retroactive since entanglement is a nonlocal phenomenon, and so can be atemporal. Thus, consciousness always becomes an entangled component within a multi-level regression comprising the superpositioned pre-particles, the experimental equipment, the computer, and the minds of the experimenter and any observer. The resultant outcome will always reflect the mutually informed relationship of the entangled particles, as consistently demonstrated by such experiments. This idea of extending the effects of entanglement to include an observer can be further compounded to include another observer observing the first and so on, becoming an infinite regression of entanglements and a veritable network of correlated outcomes. Indeed, such an entangled web of minds and matter may be how existence works.

Scientists have always refrained from anthropomorphic assertions, such as saying that the entangled, superposed pre-particles "know" each other's state. To avoid implying some form of conscious "knowing" on the part of particles, an expression such as "mutually informed" is preferred. One conceptual model, designed to exclude any role for consciousness, uses String Theory, which speculates that subatomic particles in spacetime are the ends of vibrating one-dimensional strings whose lengths reside in extra-dimensions. Thus, entangled particles are just two ends of the same microscopic string, which are always in touch with the state of each other through the body length of the string.

The nonlocal nature of their entanglement results from the string's body length being beyond and independent of spacetime. Experimental evidence confirms that an entangled particle-pair in a state of indeterminable superposition always appears to share or determine complete mutual information about each other's dynamic state, independent of any spatial or temporal separation – the principle of nonlocality. If one particle's wavefunction collapses due to observation, its state becomes manifest. Its distant twin's complementary state is then always deducible in advance of, and independent of any direct confirming detection.

Role of Mind

An ensuing notion is that perhaps the entanglement effect of apparent mutual state-awareness between inanimate objects such as pre-particles might not be what it seems. It might be that the superpositioned particles are not actually in a direct state of mutual path informing, it just appears that way to the observer. Instead, it could be that the status information link is not directly between the entangled pre-particles at all, but a multilateral entanglement between the observer's nonconscious and each pre-particle separately. If the observer were unaware of their nonconscious influence on such an experiment, it would just appear as though the particles must be communicating their outcome-path directly to each other.

Such an effect might even be a fundamental part of the construction of our reality – the melding of mind and matter, nonconsciousness and wavefunction. At the point when consciousness becomes curious about a pre-particle's eventual post-collapsed state, the nonconscious entangles with, and informs itself of the pre-particle's permutations of possible paths, with the added potential for biasing the trajectory of the probable outcome in the process. The entangled pre-particles would then collapse out of their superpositioned potential states influenced by the mentally favoured likely outcome, and still be in their correlated state of manifestation.

Although we are discussing the entangled state of mind and pre-particle wavefunctions here, there is another possible aspect of entanglement, less relevant to this topic, but which may have accumulated more associated anecdotal evidence. That is the entanglement of mind with mind, which we briefly referred to earlier as a "resonance" between individual minds and presumed to be by way of nonlocal mutual sharing between each nonconscious. Possibly, such instances as empathy, telepathy, and crowd hysteria provide circumstantial evidence of entanglement between minds. Limited scientific research into this subject has been undertaken – some funded by the military. However, the inconsistency of the evidence points to an indeterminate phenomenon. Anecdotal evidence shows up in reports of twins separated at birth leading similar lives, making the same choices, or becoming aware of each other's trauma, and feeling each other's emotional extremes of pain or bliss. Quite frequent reports arise of various forms of extra-sensory perception occurring between individuals, particularly family members, life-partners and the emotionally close, and of the phenomena of unrelated people performing altruistic heroic acts.

Partially successful remote-viewing trials – the military's chief interest – and the possibility of clairvoyance are among similar accounts. Earlier, we visited the notion of emotion as being a base "language" connecting nonconscious minds and Sentience. Mind-to-mind entanglement might present as being aware of mutually shareable or referenced information, perhaps by way of such emotion-based language. Arguably, any syntax based on emotion is unlikely to form a precise description of shared information, and so would need subjectively converting into conceptual symbols, words, or pictures for sharing. This built-in vagueness might explain the marginal hit rates in the relatively few scientifically controlled experiments reported.

We noted earlier that psyche or spirit might focus onto the individual's nonconscious mind, in the manner of a dynamic holographic-like image or an attenuated duplicate of universally distributed Sentience. One could imagine the individual's psyche or spirit as also universally distributed, with just the most substantial focus node at the spacetime location of the host individual. Then one might consider all individuals' nonlocal, universally distributed psychic fields as layered or superimposed upon one another. Thus, there could be innumerable possible contact points across the universe with every other distributed nonlocal sentient presence, though the strongest would correlate with the emotional "adjacency" of the focus nodes.

This model describes a massively interconnected universally distributed network of interdependent psyches of sentient beings, with each one forming an independent focus node within the nonconscious processes of their host. In this case, the issue of exchanging mutually shared information would not be so much one of selective entanglement, since everyone's nonlocal part of their nonconscious would already be entangled to some degree, but more like one of selective pairing by mutual recognition. Like entering a unique number into a telephone exchange, one psyche could establish an emotion-based resonance with another by choosing the right emotionally synchronic path to correspond with the target focus node. Furthermore, a familial or emotionally close association might provide enough "adjacency" to effectively accentuate such a link.

To extend this thought further, one could envisage universal Sentience itself consisting entirely of such a layered holographic network of entangled, superimposed, nonlocal interfaces with the nonconscious processes of all sentient beings. Conceptually, there might be no need to imagine a single overarching identity, such as Sentience, except as a collective noun. However, although this network model interpretation has interesting potential and shares similarities with Jung's collective unconscious concept, for the time being, it may be easier for discussion purposes to retain the initial single identity concept of Sentience.

Mediating Change

Central to this discourse has been the physicist's "skeleton in the closet" aspect of consciousness-mediated changes to the probable end-state of a pre-particle. The focus of the discussion has been on particles because experimental evidence suggests this happens. The most reasonable though controversial explanation to date seems to be that conscious involvement in the act of observation of a quantum state, directly or indirectly, may modify the likely behaviour of the associated quantum event. The concept of mind entangling with particles and indeed with other minds, as has been suggested here, is quite speculative, and there is a scarcity of repeatable experiments, thus relying more on statistical interpretations.

Consciousness-affected particle manifestation is a challenge to prove. Still, the key to evidencing a consciousness-mediated reality might be less about the individual particle-by-particle influence and more related to the impact of consciousness on what we consider as the natural laws of physics – how groups of particles might interact universally. Apart from the laws of conservation, other such fundamental physical laws include the group of improbable values of critical constants, parameters, and ratios of nature. Remarkably, many of these existing physical laws seem to deny a reasonable likelihood of being. They appear so improbable and critically fine-tuned to give us the only reality we know that minor deviations from their values of less than one percent in some cases would render the universe untenable. Nevertheless, those fine-tuning laws did evolve and establish themselves in the form of the natural physics we now observe; and the resulting evolution of energy-matter did follow a path conducive to supporting sentient life, of which we are a part.

The problem with this aspect is that although evidence for the subtle conscious mediation of the laws of physics might emerge from a search for statistical significance in a sample size of many entities, as far as universes go; we have a sample size of one. Thus, we can only infer that the probability of a particular physical outcome may have been indirectly mediated by sentience from the otherwise extremely unlikely possibility of such fine-tuning occurring by chance alone. If we could look at many universes, we might conclude that ours was the only one – or one of many – where the laws of physics supported life and sentience, but we cannot do that. This kind of circumstantial evidence from a sample size of one is an unsatisfactory basis for science to classify either possibility – random or influenced, natural or unnatural – as anything more than speculation.

Furthermore, as discussed earlier, the chances of our being in a fine-tuned universe seemingly for the benefit of the physical stability of the cosmos, life, and sentience, could result – though equally as speculatively – from an extension of the "many-worlds" concept. In this case, presuming an infinity of alternate universes, one or several like ours might well support life and sentience; and we happen to be in one of them. Similar speculative explanations might involve infinite numbers of regions, bubbles, or patches of different physics within a multiverse, of which a few – including this one – turn out, randomly and naturally, to support life and sentient beings who then evolve to question why.

While the idea of there being at least one other universe was suggested earlier, the extension of that notion to infinite numbers is intuitively questionable, and the energy requirements stupendous. Again, one wonders if the many-worlds and multiverse concepts prevail only to avoid the scientifically unpalatable alternative explanation involving participatory conscious mediation, which may turn out to be a fundamental phenomenon. A compromise solution to the credibility issue of multiple universes was raised earlier, whereby those possible alternate universes might remain in superposition as an unmanifest potential state. There, they would await the browsing light of consciousness to briefly slice into them, bringing about a moment of coherent alternative reality, before moving on to other possibilities.

In that case, the structure of evolution itself would be more a history of pseudo-random quantum wavefunction-collapsing events selecting for sentient survival and evolution, as mediated by the gaze of participatory consciousness. That would characterize Sentience as selecting from all potential in superposition, a string of temporary realities to manifest, which accumulate sequentially to form the historical path of existence as we know it, and in so doing provides us with the illusion of time. If that solution were reasonable, it is dependent on intervention by consciousness to manifest reality. The conclusion here is that a form of consciousness may indeed mediate and select for a sentient-beneficial form of existence, though the mechanism for such mediation remains obscure.

Nevertheless, given the conjecture that consciousness mediates reality at the quantum-event level of pre-particles, we might attempt to imagine a possible analogous mechanism that could serve as a model for what might occur at the quantum level to produce such macro-scale effects. Such a model would almost certainly not represent the actual mechanism; however, it may at least imply that some form of mechanism for consciousness to influence the initial formation of fundamental physics during the formative stage of the birth of the universe might be possible.

Quantum Amplification

Awareness and intent, by way of the nonconscious, may collapse the wavefunction of a pre-particle in superposition. Not only that, but it may influence the state into which it manifests. Furthermore, such an interaction can be independent of both distance and time.

However, such interactions would only occur at the quantum micro-level of physics, and thus on the face of it may appear unlikely to have much measurable effect on any macro system containing those particles, without some way of leveraging, multiplying, synchronizing, or cumulatively combining the impact of those interactions. So, let us explore how that might occur.

For nonconsciousness-induced quantum state changes to affect reality, they would need some intensifying process whereby small, seemingly insignificant quantum-level events could lead to greatly exaggerated macro-scale physical effects. For brevity, we can call that speculative process "quantum amplification." As far as i know, that is not a recognized technical term but serves here to label an unknown possible mechanism.

Though we may not comprehend such a process, quantum entanglement is likely to play a large part in it, and we can identify natural analogous models that could act as a stand-in model for the process. These model the collective behaviour of whole systems that already exist. Computer simulations of these processes are being developed to help understand forms of difficult-to-predict behaviour that fall under the overarching generic term of Complex Systems. These include but are not limited to the cascade effect, chaos theory, non-linear dynamics, fractals, statistical physics, and information theory.

The cascade effect model – like a chain reaction, avalanche, or domino effect – can be illustrated simply at the subatomic level by nuclear fission. At its most basic level, in a troublingly straightforward process, nuclear fission occurs when unstable atoms in a critical mass of purified radioactive-rich substance spontaneously release multiple energetic subatomic particles as radiation. In a dense, pure, radioactive source of critical mass – the mass at which the fission effect becomes self-sustaining – these energetic particles collide with other atoms, causing each of those to release additional particles. This new wave of radiation triggers the release of further energetic particles from adjacent atoms, and so on. These energetic particles also radiate intense amounts of heat, thereby thermally hastening the reaction. Unless moderated, the energy released follows an aggressive exponential cascade, producing a thermonuclear explosion. Conventional power-generating nuclear reactors typically control such a reaction with radiation-absorbing liquid and dense metal rods, to keep the core below its critical state.

A less dramatic model for leveraging up from quantum levels to macro-scale comes from another class of natural mechanisms – non-linear systems. Non-linear processes are, at their heart, extraordinarily sensitive to minute changes in their initial conditions. In these types of processes, minimal changes in the initial conditions of a complex system can lead to exaggerated and often unpredictable outcomes. If such changes in initial conditions were to result from conscious-mediated quantum-level particle behaviours, the resulting physical states of the affected systems might respond quite dramatically to such marginal influence.

A familiar example, actually a metaphor, of such a complex system is found in Chaos theory and illustrates these phenomena. It is the classic "Butterfly Effect" metaphor, whereby a fictional butterfly flapping its wings in, say, Hong Kong, ultimately causes a tornado in Texas, or some similar assertion involving different parts of the world. The weather, a worldwide non-linear complex system, is a notoriously difficult system to predict, even at a gross level and in the near term. The theme of the Butterfly Effect supposes that minor vortices in the surrounding unstable air caused by the butterfly flapping its wings, gradually build up into turbulence in the local atmospheric conditions. Since all planetary weather is essentially one complex system, this modified turbulence migrates and builds as the weather system it is in crosses the ocean. It then acts as a trigger event causing a tornado in an already stressed and unstable weather system in some faraway place. This example is of course, conceptual; no one has conclusively evidenced any such cause-and-effect chain reflecting this specific possibility. However, while not to be taken literally, it is a useful metaphor for describing the powerful leveraging effect of initial conditions in complex systems.

Influencing Matter

Knowing something about the feasibility of any kind of possible mechanism for amplifying quantum effects will help us appreciate how intent might affect matter. Regardless of the actual mechanisms involved, we can imagine that a process such as quantum amplification might operate similarly to examples of such mechanisms found in natural complex systems. We could imagine that in such a vastly complex system as the universe, quantum entanglement effects could accumulate and influence all histories relating to the chain of quantum state events that produce the physics we know. That would allow for the fine-tuning of key physical states, laws, and relationships into new and sentient-beneficial default values, which those eventual sentient beings would then observe as natural physical laws and constants. A dynamic modulating process such as this may even be necessary for stabilizing the ongoing operation of an evolving cosmological reality. Thus, nonconscious influences on the initial conditions of complex systems at the quantum level may ultimately be responsible for the effective working of our life-nurturing universe.

We can now consider how such a hybrid system might operate.

COSMIC INFLUENCE

Power of Intent

Experimental investigation and modelling of non-linear, complex, and infinitely complicated fractal-like systems lends support to the view that natural mechanisms may enable critical quantum-level effects to change the initial conditions of complex and chaotic macro systems. The realization that nature appears to make use of similar processes tends to support this view. Conscious awareness and coherent intentions could shape the probability spectrum within quantum wavefunctions before their collapse, and thus their likely outcomes. Once collapsed, these events can participate in the initial conditions of non-linear complex processes that in turn could fundamentally influence their development. Such initial condition amendments could then instigate synchronized behaviours in much larger complex systems fundamental to the physics of our reality. That suggests an analogous route by which conscious intent might ultimately effect changes in reality.

Undoubtedly, the nonconscious amendment of our reality is not such a straightforward process as just outlined here. The criteria of coherent intent may only come from relatively lucid minds, preferably in unison. Most minds, human ones at least, have other, counter-productive things going on within them – mostly, egocentrically oriented noise and conflicted waffling intentions. That is why positive co-creation of our existence by a materialistically dominated society may be depending on a dismally small – and unaware – fraction of the population. I am thinking particularly of enquirers, those deep-thinking individuals exploring the mechanics of existence, motivated scientifically, spiritually, philosophically, psychologically and by other fields of endeavour, curious about what is and what can be, and with the beneficial outcome for human society in mind. Our very existence may be contingent on the realizations of those meagre few altruistically motivated, authentic individuals – a rather small sample size of dedicated souls. One might then imagine that more would be better, which raises the question of how many such individuals might it take to make a significant difference. What critical proportion of society would need to adopt such awareness and purity of intent, as to effect a sea-change in the evolutionary direction of that society as a whole?

A couple of decades ago, experimenters investigated the power-escalation effects of synchronizing multiple beams of coherent laser light. They concluded that the critical number of pilot laser beams required to coax the power of a larger population of laser beams into synchronism was of the order of the square root of one percent of the total number of beams. These results were later seized on, perhaps somewhat naively, by some social-science-oriented organizations as being indicative of the portion of a human population needed to influence the whole society. A large-scale test conducted across several large U.S. cities had the goal of causing a reduction in the overall crime rate by meditation. The test involved bringing in several thousand skilled meditators from around the world, representing in number the square root of one percent of the population of the cities. Their goal was to meditatively focus on dropping the local crime rate but not concern themselves with how to do that, just on decreasing it. Protagonists for the experiment claimed a successful outcome, with a statistically significant reduction in crime over the test period of several weeks. Critics declared the results inconclusive. I am skeptical of ascribing to complex human psychological and social behaviour the same technical characteristics as laser beams, or the expectations of resolving society's moral compass issues by reference to such a simple, single-variable formula. However, directionally at least, the idea is interesting, although intuitively, perhaps one might imagine a more robust proportion of the population of five to ten percent would be a more practical core size needed to sway the whole.

Nevertheless, this idea does suggest a possible focus for proactively influencing the moral and conflict-resolution choices of human society. By encouraging pockets of enlightened, altruistic, mindful, and dedicated individuals on a global basis, we might be able to breathe new directions of mindfulness, peace, and responsible, mature, and less materialistic behaviour into this bickering race of human beings. Unintentionally, perhaps we already are. The pursuit of such a social impact might be an ideal term of reference for a spiritual organization or philosophical movement. Furthermore, in terms of the physics of our existence, many dedicated enquirers diligently pursuing discovery may already be unconsciously helping frame our reality in this way.

An interesting thing about this form of influence is that these individuals need not be in positions of conventional hierarchical power, since theirs is the power of intention, of which they may not even be cognizant, and it is not externally derived, granted, supported, or visible, but comes from within. Such a concept of power may cause global consternation, suspicion, and resistance among the more conventional political power-mongers. The implications of such influential sentient consciousness in our current society may seem colossal. However, if the influencing of quantum events were achievable by all instances of sentience everywhere, the implications would reach beyond our world, beyond our time, and point to universal co-creation.

Sentient awareness may influence the choice of physics of those not-yet-existent quantum pre-particle manifestation events – what state they will be in when interacting with our reality. However, evidence of influencing pre-particles through quantum probability biasing is sparse. Some years ago, instances of mentally regulating random number generators appeared sporadically during psychokinesis experiments conducted at Princeton University Engineering. In those experiments, an electronic random number generator, of the sort that drives the national lottery programs, appeared to be marginally, though statistically significantly, influenced by the mental effort of the test subjects. Typically, in such experiments, the recorded influence is frustratingly minor and unpredictable, though cumulatively it can have statistical validity. The results of these types of experiments are not earth-shattering proof of such effects, although neither could they be dismissed as random fluctuations.

As an analogous example, the ubiquitous transistor makes use of quantum probability biasing effects by modifying the probability of electrons being able to travel through the device. The phenomenon responsible – quantum tunnelling – describes an electron's wavefunction disappearing from one side of an insulative barrier and reappearing on the other side. Although the unbiased probability of electrons travelling through the insulation is negligible, the introduction of a small bias electric charge into the semiconductor device skews the wavefunction probabilities of the main source of electron pre-particles, causing them to most likely collapse and manifest on the other side of the insulator, thus seeming to flow across it. This biasing effect appeared to enable the main electric current to "tunnel" through the insulative barrier at vastly higher rates than the modest applied bias current – an example of cascade amplification.

That process is similar to the quantum amplification effects discussed earlier, except that those appear influenced not by an electric current, but by nonconscious intent magnified through the auspices of the Sentience network. Nevertheless, this concept may model how conscious intent could change the outputs of random number generators and even the universe by the subtle biasing of probability-driven wavefunction-collapse outcomes. Accumulated over countless such micro-events involving this form of quantum amplification, such subtle biasing could statistically give rise to pivotal initial events and values. They could adjust the evolving direction of the universe and fine-tune all associated support mechanisms and structures – effectively creating the newly "discovered" laws of physics.

It would not be necessary for sentient beings to know how to effect such changes to a complex system's outcome to bring about the required effect. All that is needed is to provide elementary experiential feedback, just as a simple thermostat informs a much more complicated air-conditioning system that it is too hot or too cold, to which the complex system then responds. The universal network of consciousness, Sentience, could then respond at a fundamental level to such systemic existential feedback from sentient beings, communicated via their nonconscious processes.

To illustrate this further, we can use the woodcarver metaphor. The woodcarver represents the collective Sentience, the wood the physics of the universe, and the tools she uses the sentient beings. From a smart tool's limited perspective, it is routinely slicing its way through the grain of the wood, providing continuous tactile feedback to the carver, although being unaware of the strategic intent contemplated by the carver. Suppose the tool comes across an inconsistency such as a knot in the wood, which conflicts with its expected responses of the material. This discovery lies beyond the tool's previously experienced physics of the wood. The feedback the tool gives to the carver might be an unfamiliar feel in the normal cutting process, excessive roughness, or refusal.

On receiving this sensory feedback, the carver can modify her response by considering many possible solutions. These may range anywhere from sharpening the tool, choosing another tool, modifying the tool action, pushing harder, or even changing the design of the carving to use that inconsistency as a feature. The carver has a global perspective to understand the many solutions to the issue, and to choose the optimum means of achieving a satisfactory resolution, which the tool does not. To the tool, the detail of the problem and the subsequent solution may be incomprehensible, though the solution instigated by the carver would seem natural.

Hierarchy of Influence

Influencing the creation of form can take place in two principle hierarchical directions. Simplified, these are the classic top-down and the bottom-up approaches. An alternative perspective might be the broad-brush of impressionism or the detailed realism styles of art. These reflect the hierarchical creative processes used by humanity in most of its pursuits. However, when creating a woodcarving or even a universe, quite pragmatic natural rules would govern the process.

The chief rule in manifesting any creation is that the ideal of perfection is unattainable. Within the fine detail of a carving, no matter how fine it is, it can never perfectly reproduce the same characteristics found in the original, or in the sentient mind's eye. As with any creation, there is always an element of approximation and compromise involved in defining the detail, which leaves room for potential future improvement. The artist and the beholder alike usually understand such approximation to be a pragmatically acceptable limitation. Since achievable detail can never be perfect and can always be improved, the whole creation will either remain inherently incomplete or be subject to a dynamic, incremental evolutionary improvement process.

One can only imagine that these natural constraints on the process of creation would be similar when constituting a universe. A dynamic broad-brush image could be invoked portraying an expanding, evolving cosmos, the provenance of star systems and galaxies. Such an image may be satisfactorily cosmos-like when viewed from a distance. However, to be realistic, the creation needs to hold up to conscious experience, detailed examination, and persistent enquiry by any sentient beings around to observe it. Functional detail and supportive history need incorporation into and throughout the broader evolving image. To provide a sufficiency of reality, the level of detail must be of finer resolution than the ability of those sentient beings to distinguish. Fine tools, effective feedback, creative imagination, and continuous improvement are required to provide that level of detail. Part of that evolving detail would be the creation of an adequate chain of authentic causes consistently linking to the overall perceived effect – its historical record. We call that physics.

Sufficiency of Reality

In physics and cosmology, the Anthropic Principle is a principle asserting that observations of the physical Universe must be compatible with the life observing it. The principle can be divided into "weak" and "strong" forms. The strong anthropic principle claims that since we sentient beings are here to question why we are here; the universe must have evolved – though not necessarily having been created – to support humanity's existence. The weak anthropic principle counterclaims that we are here, questioning our existence because non-directed random circumstances in a naturally evolving universe eventually led to the conditions in which sentience could emerge to ask such questions. Religion generally represents the strong form, physicalist science the weak form. Neither point of view is entirely correct, yet each has some validity. So just what is the connection with the frontiers of science?

As noted earlier, because of the effects of the limited speed of light, we inhabit a spherical portion of a universe – or multiverse – of unknown overall size and shape, which, although expanding in size, from our perspective is diminishing in observable content. The universe we can observe – a virtual sphere, with us at the centre – is all we can hope to interact with physically. The horizon of this sphere is determined by the inability of light from receding objects at its outer edges to travel toward us faster than the speed of their recession away from us due to the expansion of the universe. In other words, beyond that horizon, the light image of receding distant objects just cannot get here. What lies or moves beyond that spherical horizon is forever lost to us.

The classical assumption has been that there can be no faster-than-light interaction between our observable part of the universe and anything lying outside it, so it is irrelevant to our reality. This assumption is now questionable since discovering that through the principle of nonlocality, lightspeed may not limit all forms of influence upon us. Furthermore, extradimensional possibilities used to be considered mere mathematical inventions, of interest to science-fiction writers though irrelevant in practice. However, that assumption is challenged by recent efforts of string theorists and others pointing to potential extradimensional underpinnings of our physical reality. The possibilities of additional dimensions are now taken seriously, and there is recognition that such things we cannot sense, including dark energy and dark matter, can indeed affect us.

An increasingly discomforting issue for science undermines the idealized classic concept of objectivity. Until the advent of quantum physics, the subjective consciousness of the observer was assumed irrelevant and excludable as far as the methodology of an ideal objective scientific experiment was concerned. Quantum physics now suggests that consciousness is not only relevant but cannot be prevented from influencing the performance of an experiment. Despite the evidence from quantum physics experiments, many devotees of materialist science would still deny the involvement of the observer – or even the existence of consciousness itself – in influencing the results of an experiment. Their perspective is that even if consciousness existed, it is a local emergent behavioural property based on biological electrochemical activities within the neurons of the brain and cannot affect external physics. The contrary philosophical question that arises is that if there were no conscious observer present to extract information from a potential event – by any means – then might that event remain as just one of all possibilities, never to become a reality? In a generic sense, perhaps the degree of material reality relates to the degree of information relevant to, and extracted by sentience, and is only sufficient to support what that sentience experiences.

We noted the "sufficiency of reality" conjecture earlier. The reality of existence presents itself as our consciousness examines it through one or more of our senses, including that of thought – it is experiential. As we become aware of its various aspects, we develop mental conceptual models or symbols to represent the form of the experience, which enable us to conceptualize, mentally manipulate, and communicate our relationship with it. Hundreds of millennia ago, sentient beings, some being our ancestors, would have looked up, observed the tiny points of light in the night sky, and wondered. I am thinking primarily of humankind, though i want to keep the perspective broader. One might imagine that some curiosity existed millions of years before humans.

Those early animals might have at least taken a passing nighttime interest in any unusual encounters they may have witnessed. However, they probably would not have been interested enough in the individual stars to be even aware of them. To them, the night sky was effectively irrelevant, unless supporting the presence of nocturnal predator or prey. On planet Earth, it would only be as recently as fifty thousand years ago that some of those sentient beings – humankind – could hope to discuss and portray their wonderment over those twinkling lights in the sky, thus starting the process of philosophical science and cosmology.

Only a few centuries ago, we conceived of the heavens as being a fixed light-studded hemisphere where the gods lived, surrounding the flat Earth. For most circumstances of the time, that model was a sufficient reality, fit for living in that era. Subsequently, though, our curiosity became aroused. The irrelevant became relevant. The relentless pressure of essential survival-related activities relaxed, and time became available to peruse more than predator, prey, and procreation. Progress continued in our physical and mental capabilities. The reach of our senses extended, magnified by the tools we made, the sensors we created, the observations we took. Our conceptual abilities improved, along with our newfound creative curiosity. Just a dozen or so human generations ago, we found ourselves probing the heavens with simple telescopes. We discovered that we were not the centre of a shell of twinkling lights and challenged the conventional views of religion and philosophy. Gradually, we adopted scientific methods and built better models, ones that successfully matched observation, and correctly predicted earthly and astronomical events such as seasons, weather, moon phases, eclipses, and planetary orbits. This process took place in all realms of human interaction with reality. The examples of exploration of the heavens and resulting sciences of cosmology and theoretical physics are just one small aspect of life being used here as a simile for all the rest. The process of learning, comprehension, and conceptual incorporation follows the iterative process of explore, discover, consolidate findings, revise, test, explore again, and so on. This continuous, iterative process of enquiry, exploration, and integration is how reality and individuals evolve – both mentally and physically.

Work in Progress

There has always been at least one crucial implicit assumption in this learning and discovering process that needs challenging, which relates to the principle of local realism – a principle that things already exist before observation and are only locally influenced. This neglected, but cardinal assumption is that the mysterious physics of existence in all its myriad aspects is out there already established, completed, and perfectly assembled, awaiting revelation by our enquiring minds, our tools of discovery and our powers of comprehension. The controversial conjecture raised here is that this assumption is flawed. Our reality is not complete but completing, not perfect but perfecting, not awaiting us but requiring us, not being revealed but being created. The conjecture offered here is that existence is a work in progress, *necessitating* sentient input.

The fundamental proposition is that when we comprehend a discovered aspect of existence, we relate to it, accept it, and develop an intricate conceptual model of it – mental, representational, mathematical, digital, or physical – to aid in comprehending and manipulating it. Our model must stand the test of reality checks to be a credible analogue. It must be adopted as representing the new reality by others – by many – and must conform to known structural, physical laws, or show those understandings to be inadequate. Then, it is to that degree of detail, and that degree only, that the discovered aspect – the object of interest – effectively acquires reality; it becomes relevant and sufficiently manifested to our consciousness. Before the moment of conscious extraction, interaction, and comprehension of information about it, that aspect was a broad-brush, nonspecific representation of what it might potentially be; it was like a wavefunction. The discovered object of interest follows the same path as inventions everywhere – recognition, enquiry, concept, proof of concept, reality.

The conjecture that our reality is essentially inventing itself, through the agency of sentient consciousness, might be somewhat disquieting. Right now, you are probably thinking of reasons why this might not be so. One of those reasons might be the issue of event timing – how sentient beings such as us could influence formative events that occurred before life evolved. That might be a valid point, though only from the conventional perspective of assuming sentience as being emergent from matter.

When recalling a particularly vivid dream, frequently, only the essentials of scenery are present as a backdrop to the principal characters or action. Even those characters remain relatively poorly formed until called upon to contribute sufficiently credible detail to the dream storyline. The dream creation process seems to save its energy for detailing only the essential active parts. Peripheral parts remain barely adequate, just enough to provide some confusing context. It appears that the necessary detail emerges from the degree of focus of one's sleeping awareness. Supposing the potential of the countless pre-particles in the universe were available to contribute to realizing our existential dream, and that the dream is merely a creative thought experiment within the distributed mind of Sentience. The pre-particles would be like actors waiting off-stage in their wavefunction, away from direct attention until needed. Those unrealized pre-particles remain in an unmanifest state of ethereal grace – mere possibilities until required for real interaction. Then, in the presence of sentient awareness, the possibilities become probability driven. As each particle's quantum wavefunction collapses to manifest into observed existence, it becomes a certainty participating fully in the physical process of our reality.

For reality to be what it is, many fine-tuning influences would have had to be exerted on the evolving universe. Among the most critical of those influences would be those in the distant past before the appearance of humankind or any other kind of sentient being. Some of these influences would have been present from, or paradoxically, even before the beginning of the universe. However, that does not mean an explanation has to invoke a mysterious deity as the creator. Instead, we could identify nonlocal quantum effects and the extended influence of an extradimensional field of networked consciousness as the instigator of the seed of the universe. The complete physical infrastructure of the universe – the entire spacetime continuum – would reside within nonlocal Sentience, so the breadth of influence is not limited to Now, as it appears to be for us. Retroactive influencing would enable the development and fine-tuning of past structural physics of the universe, biasing its complex evolution toward encouraging its present-day structure, life, and sentience.

To illustrate this metaphorically, consider our woodcarver who, while working on the neck of the duck she is carving, realizes that for this posture, the already-carved back, representing the temporal axis – time – needs more work. She can return to the back of the carving to fine-tune the whole model. She is not limited to working on the sculpture from one end to the other in a constrained linear fashion. The carver will move all around the model iteratively until the whole is brought into a delicate balance. That is what timeless Sentience can do with its thoughtful manifestation of the universe, which lies within its realm. Sentience is unconstrained by the linear limitations of spacetime. It can experience Now and use those learnings to influence what took place in the past, even the very beginning (we will see how that might be done in the next chapter). That version then causes a revised sentient conscious awareness of Now to be manifest. Just as the woodcarver progresses the duck, so does Sentience-mediated evolution iteratively tweak and tune the emerging spacetime universe.

However, even for timeless Sentience, the future is not definable ahead of time; it remains probabilistic. It cannot be pre-ordained; it has yet to happen. The future may sometimes seem relatively predictable, based on high probabilities. Nevertheless, in the physical world the future's actual manifestation is heavily influenced by the present, by the uncertainty of quantum jitters, and on the part of sentient beings, by nonconsciously sponsored free will. The future remains a world of potential possibilities and probabilities, though nothing yet committed and condensed into certainty, for then it would be manifest.

The woodcarver cannot start with a completed wooden duck; she can only work through the development of the duck model. At some future completion point, through the limiting nature of the wood, the tools, or her skills, and despite all her careful fine-tuning and tweaking, she may not – most likely will not – achieve exactly what she had initially envisaged. The carving might be like the duck she had first contemplated; it might be a better, worse, or different rendering of her original thought. Future detail cannot be preordained until it becomes manifest as Now. Likewise, Sentience can amend the present physics of the universe, and retroactively the historical past, but not into the future. For, if Sentience were to think the universe into a certain future, then that would become the new Now – and maybe it just did!

Our nonconscious link of psyche or spirit is a nonlocal, individual window onto Sentience – an interactive holographic representation of it, and an access point into it. It represents a user interface. When we have enough information to realize – when we understand correctly, coherently, rigorously and without a doubt – how something can function, how something manages to exist, then that feedback is what Sentience can integrate into its global manifestation as detail. This detail, which forms the physics needed to make something real and substantive, must be at least as sufficiently real as our sentient awareness, our conscious comprehension, and our sensory perception of reality allows and requires of it. The material reality of our existence, the actual manifestation of physical detail – each dotted "i" and crossed "t"; the very limit of textural resolution of our reality – is determined by what Sentience has perceived from our experiential feedback. Otherwise, it remains unmanifest, a backdrop illusion of probabilistic possibilities. Through this mechanism, all we sentient beings create the detail of our collective reality, co-creating the formative physics of this virtual universe and our very existence in it. We are writing ourselves into the drama, painting ourselves into the picture.

Communion and Co-creation

"I doubt, therefore i think, therefore i am" ("Dubito, ergo cogito, ergo sum") is the famous philosophical statement of Rene Descartes. Later abbreviated to "I think, therefore i am," it is a phrase both figuratively and literally true, and in this context might be even more profound than its author had intended, having transcended in meaning from existentialism to co-creationism.

The power of conscious, rational thinking in humankind generally arises in the neocortex part of the brain system. As noted earlier, this part of the brain has bloomed during the last hundred thousand years or so, an extremely recent event in the overall timescale of terrestrial life. So often, we think of the body as the dominant entity and the skin as the limiting envelope, an individual's entire boundary. With such an egocentric, materialistic paradigm, we talk of "my soul" – well, some do – in the same possessive sense as "my finger," "my mind," or "my dog." To get past the limitations of this restrictive paradigm, we need to comprehend and live from the perspective that the body is the impermanent appendage to a more abstract form of self, which in turn, is a link into a timeless universal network of collective nonconscious. Our body is the temporary party and is the transient focus of mind, which, through our nonconscious, is our interface with Sentience. While Sentience is timeless, the mind of a sentient being is limited to the temporal leading edge of existence – the indefinable, wafer-thin, sequential cross-sections of changes forming the ever-expanding outer boundary of accumulated local existential history. This boundary we call Now.

It is not the individual's consciousness that interfaces with universal Sentience, but the nonlocal part of the unconscious mind – the nonconscious. The nonconscious is also where heart- and gut-centred emotions are rendered, from where instinct and faint extra-sensory perceptions emanate for conscious articulation, and where the psyche, the focus of spirit, is centred. Although we are not aware of communion with Sentience – most of the time anyway – this nonconscious dialogue with Sentience is more fertile than one might imagine, and more than one could comprehend. Ongoing in real-time and always on, it is dynamic and interactive, emotional and visionary, inclusive and comprehensive, compassionate and intelligent, curious and creative. Within our mind, the nonconscious is always aware of the conscious, sharing conscious thoughts within the hidden dialogue.

Except as children, our conscious minds spend most of the time engaged in discursive thinking and egocentric distracting activities. These can include self-justifying posturing to bolster self-perceived value, berating or praising oneself, hiding self-perceived inadequacies, or confirming to self a sense of individual identity and separateness. This kind of mental noise contributes little value to the ongoing nonconscious dialogue with Sentience. In moderation, some egoistic reinforcement of individuality is not a bad thing. In the West, most people like to value independent thinking, taking responsibility for self and pursuing their path through life. However, when our egocentric consciousness spends unhealthy amounts of effort, creating unnecessarily self-indulgent noise, it limits our awareness of being a participant in the universality of Sentience. It inhibits individual and societal efforts on the evolutionary journey of arising enlightenment and transcendence.

When characterizing a deliberate act of indirect communication with Sentience through meditation, it is not our day-to-day superficial thinking that is relevant, but a mental process somewhere between a meditative quietude of contemplative surrender, and calm, conscious, coherent thought. It is a place of relaxed introspection, a space for self-enquiry, profound contemplation, and emotional nakedness. This condition is achievable by all people of all philosophies and traditions provided they are open to it. Included in this state would be the workshop style of active meditation referred to earlier, some concepts of meditative prayer, and authentic, self-appraising contemplation. When the veil between the conscious and spirit grows thin enough, the conscious mind can glimpse beyond the rational, toward a wondrous unbounded interactive communion with collective Sentience. This communication process is usually not that precise, perhaps being borne of an emotional base. Still, it can develop to a point at which some knowledge of potential within Sentience is directionally available for enquiry. Deeper comprehensions may be restricted only by the limitations of the individual sentient mind.

Most of us would like to think that we would communicate with Sentience rationally, even logically. However, the more significant part of the individual's unconscious mind is responsible for manifesting, processing, and expressing emotions. The conscious part, associated with rationality, logic, and the left lobe of the neocortex, seems hardly involved at all in such emotional transactions, except perhaps tangentially, after the experiencing of the emotion. Then, once in a state of awareness of the emotion-based transmission, it may play more of a catch-up role – trying to rationalize, react, negate, control, fulfill, manage, or otherwise participate consciously within the already incurred emotive experience. An emotional communion with Sentience may exclude the individual consciousness completely, leaving it perhaps with inexplicable uplifting feelings of love or well-being. Sometimes, an emotional thought gatecrashes its way into consciousness, eliciting out-of-the-blue surges such as love or anger, without the cause of the feeling being present in the conscious mind.

There seems a strong connection between emotional energy, whether negative or positive and the infamous ego. Doubtless, the ego can grab hold of emotion when it deems its performance or storyline as threatened, compromised, or even supported, and can immediately mould it into a shield or weapon against any perceived criticism, or elevate it as a trophy. In this way, the ego could disturb any emotional transaction that our nonconscious may be conducting with Sentience. Since the ego invests heavily in the mindless, and paradoxically mind-full chatter that can flood much of our thoughts, it can interfere with coherent communion with Sentience.

Nevertheless, many positive emotional values can enable, support, contribute and result from quality communion with Sentience. Included in the syntax of such emotional dialogue would be virtues such as compassion, love, peacefulness, creativity, and all-embracing, unconditional loving-kindness. Compassion creates a basis for communion with Sentience, and it might be the main dialect of an emotion-based language employed in that dialogue. Such emotion-based qualities, when amplified by profound intent and leveraged by others of similar sentiment, may also contribute to co-creating and moderating our reality. Through an unspoken dialogue in the language of emotion, we sentient beings nonconsciously and uncomprehendingly strive toward – to use a Zen phrase – the interdependent co-origination of our existence.

Creating a Reality

In this chapter, the perspective has been offered that the universal evolutionary process is a continuous work in progress. A controversial suggestion was that the detail, the apparent solidity of our reality has no absolute, independent, intrinsic pre-existence. Instead, we sentient beings, we spirits in hard-bodies, as integral parts of Sentience, are the smart-tools that indirectly, through nonconscious connections to our awareness, curiosity, and comprehension, provide the necessary experiential sensory feedback to Sentience to enable suitable detail to happen.

The creation of this dependently arising detailed reality is an ongoing iterative process. By being curious and diligent about "discovering" the imagined detail, we cause a sufficiency of it. The arising of our emergent reality is dependent on conscious enquiry. Once created, and consistent with the whole and the history of previous versions, it becomes the appropriate level of a sufficiently functional, communal reality for all sentient participants.

Admittedly, that is as extreme a version of the anthropic principle as one could envisage. It is saying that existence is neither absolute nor inherent, but indirectly arises from, relative to, and dependent on, conscious awareness. Collectively, we sentient beings are the creative artist's smart tools, her excellent motor skills, and her keen imagination. We are that artist – and the art.

Now, we will investigate how the sort of feedback that sentient beings provide to Sentience enables creative changes to be made to the evolving physics of the universe before their existence.

AGENCY OF SENTIENCE

The Temporal Paradox

"Time was invented to prevent everything from happening at once" is a quotation attributed to many, including Ray Cummings, John Wheeler, and Albert Einstein. We touched earlier on the illusory nature of the alleged dimension of time. We noted the general awareness of time as a clock that registers the absolute temporal movement of things, or as a temporal river of happening, that flows by at a constant speed of one second per second. We examined the necessity to our reality of incessant subatomic motion, the so-called quantum jitters, being like a non-stop juggling act conferring the illusion of solidity with an arc of virtual reality. We saw that the classical notion of time is a human construct, introduced to label the perceived intervals between changes. We are in the ambivalent situation of having an excellent practical grasp of the physical impact of time without knowing much about its illusion.

If we were to consider a distributed field of undifferentiated consciousness independent of time and space, spacetime could appear as a shared emergent construct manifested within the collective nonconscious of sentient beings forming that conscious field. That concept of Sentience would then conform to the quantum principle of nonlocality, unrestricted by spacetime, hence appearing as both ubiquitous and eternal from our spacetime-limited perspective. Time would be the sequencing attribute of selected changes within its creative dream and ordering those changes linearly really does avoid the paradox of potential events in the chain of existence materializing all at once – or indeed, haphazardly. It allows an interdependent sequence of selected events to unfold by way of a linear cause-and-effect-related process at the macro scale, though less so at the quantum level.

One of the difficulties in challenging the conventional, dualistic, chronological understanding of time is that it appears to work so well in real life. Similarly, deterministic Newtonian "classical" physics has worked so well in the past and still does at a modest scale. Understandably, the new and counter-intuitive ideas of relativity and quantum physics, which were well outside the traditional mindset, had relatively slow adoption by the established scientific community during the last century. When working with our present understanding of time, various thought experiments show the inviolability of time travel to the past. One of these is the infamous "grandfather paradox," where a person goes back in time and kills his grandfather before the conception of his father. So, does the time-travelling murderer still live? Moreover, if he were to exist, how – and if not, who went back in time?

There are a couple of science-proposed solutions to this apparent paradox, which do not involve bending the universe. The first is that the paradox illustrates the impossibility of physical time travel into the past – period. The second is the "many-worlds" theory, which proposes that at each potentially paradoxical event, the entire universe, and its timeline (or worldline) splits apart, with the one result following one possible path, and the other, the other. The following example assumes only two possible outcomes from such an event, which is overly simplistic. In this case, the person in, say, timeline A – let us call him Charles – goes back in time intending to kill his grandfather. He then returns to his present where he still exists because history records that his grandfather survived an attempted murder in timeline A, where witnesses reported an unknown suspect fleeing the scene in an unfamiliar vehicle badged as a "Stealth Industries Timerider-2.0."

Meanwhile, however, an alternate timeline B formed at the potential murdering point of his grandfather, which splits off from the murderous Charles' A reality. In this B reality, Charles is never born, because his potential grandfather never procreated, having been allegedly murdered by a mysterious person who introduced himself as Charles, did the deed, then instantly disappeared without trace. So now, you have two alternative versions of the universe, version A where Charles, and his grandfather co-exist, and version B, where neither does. However, in this B-reality, there was a particular person who might have become a grandfather, but who died before he could sire children, and there never was any time-travelling Charles.

I suppose that is some kind of rational explanation – somewhat laborious, though. Frankly, it seems to be a wasteful way of getting around the temporal paradox issue. Just imagine if new alternative universes popped into existence every time something threatened any timeline with a paradox – they would be all over the place. Only a few significant bifurcating events each day and we would be knee-deep in many-worlds – hard to keep track of them! Furthermore, what would constitute a significant event? I can appreciate the murdering of a potential grandfather might be significant, but what about a flower being cut and therefore not seeding, or a single pre-particle wavefunction collapsing or not? The mind boggles at the amount of squandered energy required to install, then maintain, this ballooning infinity of alternative universes. Given these choices, for entirely pragmatic reasons, i tend to favour the first of the two solutions – no time travel.

Many-worlds Unrealized

However, i venture to propose an alternative view, adapted from the "Many-worlds" interpretation. It is one in which all possible alternative worlds, except the one consciously experienced right now, remain in a superposition state of unrealized potential. This amended many-worlds concept would comprise a patchwork quilt of overlaid possible, *but unmanifest,* alternative universes. Each one of these alternatives consisting of pre-particles in an unrealized virtual state of coherent quantum superposition – as wavefunctions. Such potential alternative universe worldlines, at every possible fork in each timeline path, do not have to become manifest, but instead, persist as insubstantial possibilities. The wavefunctions of these potential alternative many-worlds are never collapsed by conscious awareness. Instead, they remain in their unrealized states until browsed into manifest reality by our collective nonconscious. There would not be the enormous causal energy drain that would otherwise be required to maintain multiple manifested versions of every alternative universe.

Consequently, the balance of those horrendous numbers of many-world alternatives not falling under the gaze of consciousness would remain in a perpetual virtual state of potential, like dreams undreamt – just possibilities requiring no manifestation energy. That is, unless or until, coherent sentient awareness comes meandering across the potential many-worlds landscape, picking its way through the more viable of them, sampling from one to the other, and following a continuously sequenced forking path selected from alternative potential worldlines. Imagine this as a cow aimlessly browsing the greenest patches of grass in a lush, multiverse pasture of infinite possibilities. Each consciously noted blade of potential grass becomes the more probable candidate for sentient experience. It collapses into a moment of manifest reality as the bovine's grazing gaze focuses upon it, samples, experiences, and digests its information content for one glorious instant of Now. It then becomes immediately assigned to the past, a memory for the historical record. The conscious gaze makes its next sequential choice at yet another fork in its meandering existential path crisscrossing the landscape of potential – and moves on.

In this less energy-intensive conceptual metaphor, all the unmanifest many-worlds, or potential alternative multiverse worlds, whether finite or infinite in number, would be available as a potential resource inventory of possible conscious experiences for the Now. The timeline of the collective nonconscious reality, our mutually experienced manifest worldline, is mediated by nonconscious elements associated with Sentience. It would follow the consequential path from one node of Now at a potential alternative universe worldline fork, and on to the next candidate, wherever the sentient collective nonconscious caused itself to be.

As with Schrödinger's wavefunction equation in quantum physics, the observation by conscious awareness extracts status information, which then collapses each chosen slice of potential existence from the rich superpositioned content of all possible permutations of virtual universes found within its wandering existential path. The resulting freshly manifested form becomes added to the cumulative history of Nows, as awareness of it passes through that instant experience of Now.

Our conscious version of history, our sentient life path, would then be a collage of consecutive snapshots of expiring Nows selected sequentially from all possible potential alternative existences. The past would be seamlessly interwoven from a sequence of virtual, temporally unfolding, experiential, three-dimensional, interactive projections of experienced Nows. Our entire experienced "universe" would comprise the summation of consecutive histories savoured from jagged cross-sections sliced through alternative possibilities. Like some exaggerated ripple-cut potato chip, the meandering swath cuts sequentially through, and samples from, all possible potential spacetime alternatives. It connects all points that inform sentient consciousness – creating the only sure path supporting the history of sentient-perceived Nows.

Then, one might suggest, if all such alternative universes remain virtual except for each selected cross-section of collectively experienced Nows, why would one even need the concept of many-worlds at all, except to explain the time paradox? Why not just retain the concept of Sentience selecting each wavefunction-collapsing outcome that contributes to our universe from all possible choices among potential spacetime realities, thus causing the consistent manifestation of every Now-scene that we sentient beings experience? The existence that we perceive, would then comprise the sum of histories of all those selected unique Now events, which have led to our observing it.

Far from being a universe that began with a random quantum event, the Big Bang – that alleged defining moment of creation – did not start the universe; it has effectively completed it. The discovered fine details of the past are only necessary to complete the historical model of experiences leading to the present, which for our conscious awareness, is all that there is.

We have seen that, although the creation of the universe may well have resulted from the kind of quantum singularity event envisaged by science, that event represents the equivalent of an afterthought. Our collective nonconscious within Sentience browses a field of all-potential, creating our sense of experiential reality. We conscious, sentient beings are unknowingly choosing our Now from all potential possibilities, through the mutual influence of the nonlocal part of our unconscious. Our entire collective history is an accumulation of all such previous Nows.

Science is now, with exotic, verifiable, quantum-level experiments, beginning to cast doubt on the sanctity of the concept of the inviolable flow of time. This doubt has raised the technical possibility of locally generated information, albeit currently of a limited nature, such as particle status, as being available for nonlocal sharing, transmitting and even mutual influencing anywhere and at any time. That information can be shared beyond conventional temporal and spatial lightspeed transmission limitations. Scientists are reluctantly having to consider that there exists a state, perhaps only partially associated with the observed quantum-level behaviour of subatomic physics, from which spacetime is emergent. In this state, our known physical laws and causative limitations do not prevail, and we are nonconsciously sampling from it to form our existence.

To be clear, we are not discussing interactive time travel, of sending anyone or anything physically back in time to promote some reaction deemed beneficial or detrimental to our existence. This concept is not time travel per se, but a continuum of nonlocal feedback loops, enabling a past to support the present, and the present permitting and encouraging it to do so. It is about the nonlocal influencing of the probabilities of entangled potential quantum events – of particle emergence – occurring throughout the past. This adjustment of the past physics of particle interaction allows a sentient-supportive present to be experienced by us. However, each such change in existence is undetectable by us since we can never know what any alternative condition might otherwise have been.

Broadly speaking, each pre-particle wavefunction can be accumulated into one larger multiple pre-particles wavefunction of the whole. In the extreme, this accumulation process can continue until all the pre-particles in the universe can be described as one wavefunction. Thus, on a cumulative basis, these individual adjustments become statistically significant enough to affect the manifestation process at the macro scale, moulding and supporting sentient beneficial Nows throughout all earlier versions, and creating the physics we experience. Such adjustments would amend possible behaviours of physics back then and modulating their critical influence on historical events necessary for fine-tuning the Cosmos into the experienced Now.

Retroactivity

To support these assertions, we need to revisit one of the more baffling implications of nonlocality at the quantum-mechanical level. The "delayed choice quantum eraser" experiment that was noted earlier showed that John Wheeler's conjecture about the ability of a quantum effect to predetermine its past cause had merit. It endorsed the counter-intuitive notion of retroactive causation (also termed "indefinite causal order"). One of the physicists pursuing this line of research, John Cramer, termed the process as a "transactional interpretation of quantum mechanics" that effectively considers a quantum wavefunction as propagating backwards in time, as well as universally throughout space.

This interpretation affirms the ability of quantum entanglement to influence events in the past, by those of the present. On a universal scale, for example, the real-world implications of this result are such that a photon of light emitted in the past from a distant star some 13 billion light-years away might enter a local observer's conscious awareness in the present. That would effectively entangle the history of the photon with that of the observer and empower the observer to permit the photon emission event to occur the way it did, more than 13 billion years ago.

That implication suggests a thesis that could describe how nature might enable a sentient-beneficial universe to evolve. Combining this idea with the variation on the many-worlds concept just discussed, leads to a scenario whereby even within a single universe, all possible events may be in a virtual pre-particle state of superposition. They would only become real once permitted by the collapsing of the universal wavefunction through selective entanglement with Sentience, affirmed by the existential experience of contributory sentient beings.

Thus, the evolving universe acts as a complex control system whereby only those historical causal chains of virtual possibilities that become experienced by sentient beings are permitted to manifest. Sentient beings are the thermostat in the complex air-conditioning system. Conversely, virtual possibilities supporting no sentient awareness naturally receive no sentient permission to emerge from their superposition state and so never achieve reality. The existential experiences of sentient beings act like a "strange attractor" – a set of values toward which a complex fractal-like system tends to evolve toward under a wide variety of starting conditions of the system. Chaotic possibilities in the past are thus being summoned and shaped to encourage a sentient-supporting universe existing in the present. Such a conclusion substantiates the otherwise extraordinarily unlikely conditions that have led to this oddball universe evolving, at least locally, to support sentient beings – we in the present, permit the supportive past.

As more of the experienced universe entangles with Sentience throughout spacetime, and over previous generations of itself, its history becomes more substantial. It becomes more entwined and in tune with the perception of reality by its tenant sentient observers, and increasingly conforms to the coherent interpretations of reality resonating within collective Sentience. This entangled manifestation then becomes the substantial content of Now with which those sentient beings interact. Those sentient perceivers then innocently "discover" the unlikely workings of their nonconsciously co-created reality.

Manifestation

We – the larger "we," all sentient lifeforms – are the manifesting agents and voting members of that ubiquitous field of Sentience. We are the designated executive detailing agents of all that exists – Sentience incarnate at a medium scale. We – this time focusing on humankind in particular – are still relatively primitive beings, both physically and mentally, compared to what could be on some arbitrary universal scale of possibility. For a potentially advanced species, we seem quite immature; our society's median level of overt wisdom sometimes appears equivalent to that of a toddler, and we remain fully engaged with suffering, superstition, substance, sensation, and certainty.

Nonetheless, we are evolving, slowly transcending our primordial limitations – over several thousand generations to date. If only we were more consciously aware, we could already confidently count ourselves as included among the fine-finish manifestation tools available to Sentience for the evolution of existence. All sentient beings indirectly cause the adjustment and fine-tuning of the known physical parameters of the universe, so we are participating in a glorious living creation. The irony is that we remain incognizant of our creative role because it is in our collective nonconscious where our agency with Sentience is active; and so, we continue, unaware.

Once again, our artistic woodcarver serves as the analogous process of universal creation – a creator creating. We imagined earlier how the artist would rarely have a finalized replica of the finished piece in mind before examining the subject and raw material. Not only the choice of subject but also the physical characteristics of the raw material will often be a factor in moulding the mind's image and the creative process of the artist. Michelangelo was said to first gaze upon the block of marble to perceive what was within, awaiting revelation. Once the mental image of what needs carving is established and reconciled with the available medium, the process will involve some initial rough outlining of shape. We noted that an artist could not just complete the finished product with one sweep of her carving tool. Nor, for quality work, could she expect to use the same instrument for the roughing-out as she would need for etching in the fine lines of detail. The fine-tuning of the carving can only occur by repeatedly returning to partially finished work though with a lighter hand and a more delicate tool each time. Creation is a hierarchical and iterative process. Tool choice will also be hierarchical, trending toward the use of more specialized tools, the closer one gets to a detailed finish.

If creating a universe, the process would be constrained somewhat by choice of both the subject and the object of the work, by the imagined form and the range of possible physics of the potential raw material. Again, one cannot jump straight to the result – even over six days. It must be an iterative process of coarse-shaping and fine-tuning; and if dynamic, it needs to develop, evolve, and renew. If ever there were such a thing as the perfect artist, he, she, or it would still have to work within these types of constraints and follow a similar process.

Enacting this evolutionary procedure of iterative fine-tuning using fine tools is where we mid-scale beings collaborate in the process of the creation of our reality. This interpretation represents an extreme form of the anthropic principle, where anthropic means related to mankind; however, as already noted, human sentiency may be but one small facet of universally distributed Sentience. Thus, the term "anthropic" applies to the much broader context of all sentient beings. Perhaps a more inclusive word such as "sentiothropic" needs coining instead, to avoid seeming excessively human-centric.

Humankind is a curious paradox. On the one hand, we are so arrogant as to assume that we are the sole purpose of a god's creation, and the only worthy sentient beings around; and on the other hand, so humble as to imagine ourselves mere pawns on a god's already completed chessboard of a universe. When examining the naturalness question, we remarked on the improbable odds of every one of a handful – well, maybe two – of fundamental universal natural physical constants and ratios being of just the right value needed to cause and evolve this Goldilocks choice of a life-supporting universe. From intra-nuclear and electromagnetic forces to the gravitational constant, from the development of atomic structure to the miracle of carbon, there are many examples all clocking in at precisely the right value to hold the universe together stably. That stability being not only physical, but also permitting conditions for life and sentient beings, and encouraging pockets of these to evolve.

For many of those critical constant values and ratios of nature, even minor deviations from their present value would render every material thing in the universe, from atom to galaxy, at best untenable for life and at worst physically unstable. Some would deny any other agency except random chance in the selection of these constants and argue that they only seem impossibly long odds because those contemplating the issue just happened to evolve in this improbable combination of sentient-beneficial circumstances. That is the weak anthropic principle, and on a probabilistic basis alone, it does not wash.

Speculations on the reason for this universe vary. We did touch briefly on some earlier; broadly speaking, they lie in a range characterized as follows:

- Intelligent design – a creator designed the universe to support the emergence of intelligence (religious view)

- The incidental universe – the universe just happened to turn out the way it is (a random chance)

- The unique universe – a deep underlying unity in physics caused it (the Theory of Everything)

- The multiverse – out of many possible universes, we exist in one, or the only one, that allows us to evolve (a probability bet)

- The toy universe – we are living inside a virtual reality simulation (an advanced video game… or an experiment)

- The sentience-seeking universe – the Universe must evolve towards sentience, and only universes with a capacity for consciousness can exist (this is similar to John Wheeler's "Participatory Anthropic Principle" – that observers create reality)

The nature of experiences recounted in this book point more closely to this last reason and lead to the conclusion that the universe is neither accidental nor designed. It is a dynamic, fractal-like, complex system evolving within a field of unconditioned consciousness under the influence of a strange attractor. That attractor is the presence and evolution of sentience. In the process, through the mechanism of nonlocal quantum entanglement, evolving sentient beings provide existential status feedback loops to the universe system, which inform the experience of existing, allowing the modulation of the historical physics of the universe, permitting it to develop in the way that will support those same sentient beings. Conversely, any evolutionary path of physics that has not led to sentience in its future will fail to manifest due to that absence. In this respect, the evolving universe is a cooperative one, and its only realizable history must support the sentience that permits it.

The key to our existence is tantalizingly on the leading edge of physics and stubbornly reflected in the lost direction of spiritual traditions. For many, the answer may seem incomprehensible, something to be avoided or denied, something perceived as beyond the grasp of individual intelligence. However, the formation of existence is none of these things; it is a magnificent autopoietic, self-referential loop. It is *Us*.

A Brief Recap

Within the Sentience-inspired infrastructure of the universe, forms of mental holographic subsets of it are necessary – individual facets of sentient nonconsciousness. These are the smart tools that develop to engage physically and mentally with reality. They provide experiential status feedback to Sentience, thereby influencing the physics of the infrastructure toward nurturing sentient beings. In the case of something as complex as a universe, these tools, working at a fine level of physical detail, could be likened to delicate micro-surgical robotics. They would need sensory feedback, subtle motor control, and autonomous consciousnesses to guide them while sharing their experiences and the artist's immediate intent. Such smart tools would need to be fully integrated, essentially an extension of the artist, and be responsive to and a true reflection of the artist's desire. To get any physical traction, their form would have to share the same atomic structure as that being created. Therefore, they would have to consist of the same bodily material, combined with a sentient intelligence that resonates with the artist's intentions through the network-shared consciousness.

On the scale of the universe, such fine-detail tools might best be mid-scale, partway between subatomic micro and star system macro scales. As such, we are. They might come in other shapes and sizes too, though their common feature would be physicality and sentience. It is sentience that maintains the experiential feedback link back to the artist, and which reflects that artist's consciousness. These fine-detail tools may not be aware of the overall function of the work in progress, concentrating as they are on the parts closest to them.

However, by nonconsciously employing the same techniques as the artist – coherent awareness, invention, and intent – these tools, working detail by detail, unknowingly combine their efforts in contributing to the entire creation of the artist's inspiration. In the process of creating reality, acts of observation, awareness, curiosity, or discovery provide the feedback allowing the manifestation of the creation and its history in sufficient detail. They bring it into being, converting it from a broad-brush concept into a fit-for-purpose physical reality.

Temporal Feedback

We used a simple analogy to address aspects of time earlier, whereby the head-to-tail dimension of the duck represents time. To the artist, it may not matter which part of the duck she carves first, since it is an iterative process anyway, revisiting every part many times and transforming it from the original featureless block toward the mind-held projected image. The carver is free to enhance duck parts from any accessible part of the block, gradually whittling it down to reveal the final form as imagined. Every iteration essentially represents a generation change, and each change sequence, a unit of time.

We also made the pragmatic assertion that, despite religious proclamations to the contrary, the evolving universe is unlikely to be perfectly complete – ever. Neither is it likely to have been entirely initialized by a single primary deterministic event, such as the Big Bang theory and the elusive Theory of Everything would imply. Time's arrow could not wholly restrict the universe creation process to a one-time, one-shot deal with its associated highly implausible odds of success. To enable the fine-tuning of the natural constants of the universe independently of time's arrow, a dynamic continuum of subtle, iterative, temporal revisionist modifications – temporal feedback loops – would have to be an essential part of the system of creation – the carver's whittling process. Hence, sentient beings are not merely passive observers within a perfect, clockwork universe, but instead, participate as experiential feedback loops in the iterative and evolving process of fine-tuning physical existence.

The question remains: how can we sentient beings, dimensionally constrained, made of stardust, and for all practical purposes restricted by the arrow of time, participate in fine-tuning the universe independently of time's arrow? We cannot break the laws of physics and travel back in time, yet we have noted that at the quantum levels of existence, cause and effect appear reversible. How could that enable us to influence something that for us has already happened?

This important question moves us forward from a spatially oriented iterative creative process to the concept of a temporally independent one, using "temporal feedback." Temporal feedback portrays the ability to insert subtle quantum level changes anywhere along a timeline. That may be the material of science fiction perhaps but is also an insight into what may form the underlying stability of a dynamically evolving universe system. The temporal feedback concept depends on retroactive causation, a demonstrated quantum level effect whereby, while in superposition, an end-state effect appears able to influence its antecedent cause.

As a concept, temporal feedback resembles real-world control system feedback. Virtually all control systems have some form of feedback, usually negative, to enable stability and purpose through one or more automatic self-control loops. We have mentioned a simple example of a thermostat alerting the air conditioning system once the set temperature is reached. Complex control systems will make use of multiple types of feedback loops, which can include delays and phase-shifts. Feedback, as is the case with tangible forms of feedback loop such as audio feedback, where a portion of the output is inserted back into an earlier part of the input, can be critically powerful.

Feedback can be positive, causing the signal in audio systems to become unstable, resulting in that shrill scream of acoustic feedback to which most of us have been an unwilling witness. It can also be intentionally negative, causing some attenuation of the heard sound, though increasing its stability, harmonic richness, and quality. A sophisticated recording studio sound system with its various feedback controls and filters is somewhat analogous to the idea of universal temporal feedback. In audio systems, the various feedback arrangements create variances in quality and phase between the incoming sound source signal and the amplified loudspeaker output, analogous to the evolving universe "system" and an occupant's experience of it. The influence of audio feedback loops can create a whole range of desired effects aimed at enhancing the listener's experience – the equivalent to the physics of our reality nurturing life and sentient observers.

A good sound system should provide stable performance with balanced feedback, a degree of self-regulation, and tolerance for unplanned events. The audio system will usually give the best, most authentic experience for the audience if the soundboard controller is right there among the audience members witnessing the experience from within their environment – not just watching gauges on a control panel – and fine-tuning the various controls for optimal heard sound. He needs real-time feedback as one with the audience, and part of that feedback is sensing the audience's reaction to that experience.

Some parallels between the audio system metaphor and an ideal consciousness-driven universe are apparent because the chief goals and functionality are directionally similar. A universe that was not accidental, or i suppose even one that was, would have to have specific systemic characteristics to deliver an authentic, stable, existential experience to a sentient observer if they were part of the experience, or else they would not exist. The universe system would have to evolve to include environments eventually suitable for the development and support of its sentient experiencers. It would also have to be consistent and robust in its response to unpredicted change, and not be inherently unstable. Finally, it would have to have the wherewithal for fine-tuning feedback adjustments, so that its performance, and the sentient observer's experience of it, could co-evolve while retaining stability.

Ideally, then, just as the concert sound-technician and the audience should be as one, so too should universal Sentience be as one with the embodied sentient observers to enable optimization of the universe's fine-tuning feedback adjustments. Here, the "audience" would be the universe's network of executive feedback agents – all of us. Even if the universe were accidental, it would still exhibit similar sorts of characteristics to accidentally foster and sustain sentient observers such as ourselves, who were accidentally evolving within it; or they would not.

In contemplating how such a temporal feedback type of fine-tuning would work, it might be tempting to imagine a single temporal feedback insertion point where retroactive adjustments occur. That might seem to point to those sensitive initial conditions of the Big Bang as the starting point for where we are now; a point from which the life of the universe could theoretically extend, dependent on different initial conditions and results. The physicist's Holy Grail, the so-called "Theory of Everything," posits an algorithm that perfectly describes those initial conditions; thus the entire unfolded history of the universe up to Now could then be explained. This determinist-inspired dream is a fallacy. It is hard to imagine the gross inefficiency of any self-correcting complex system that would have to keep restarting from scratch to fine-tune its outcome.

The only initial conditions that we humans can directly affect are in the Now. Our sole existence is Now, and we know that how we affect Now may influence our future. However, in a spacetime-dimensioned system as complex as the universe comprising multiple networks of sub-systems, a profusion of initial conditions exists everywhere and everywhen. Thus, to be effective, major and minor temporal feedback loops must extend in all directions throughout the universe and back throughout its history to every part of it. This feedback adjustment is an unseen though systemic part of the quantum fabric supporting existence, and is a continuous though subtle process. As with the sound system feedback, it is active all the time and has the same order of resolution as the reality it supports. Thus, the Sentience-supported universe acts as a complex process control system, continually adjusting itself retroactively along its timeline to approach its goal of nurturing sentience and to evolve its existence into an uncertain future. Its very existence is dependent on its ability to modify itself iteratively through feedback – the essence of any stable control system.

A New Paradigm

In summary, the challenging concepts explored in these pages introduce a new paradigm of reality – that the present, enabled by the quantum principles of nonlocality and entanglement, influences the past. The driving force of such influence, which we call Sentience, is a universally distributed, decentralized field of undifferentiated consciousness, characterized as the unified interleaved and multilayered collective nonconscious of all sentient beings, informed and informing through that node we call psyche or spirit. Its systemic goal or attractor is to evolve and nurture sentience.

Sentient beings provide Sentience with experiential sensory feedback, prompting it to conceive, direct, and develop existence, thus reshaping the probability-determining form of pre-particle wavefunctions. As superposition states collapse by being perceived by sentient beings, the higher probability of contributing to sentient-beneficial reality is permitted, leaving all other possibilities in an unmanifest state — effectively nonexistent in a physical sense.

This sentient influence is independent of classical spatial and temporal limiting conditions. Our operational concepts of time, cause, effect, and duality have been indelibly impressed upon our minds by our experience with "reality" as we know it - the paradigms of the day. Indeed, at the macro scale, our observed reality seems to require such a sequential constraint in order to work for us, and for our informed, conscious minds to comprehend.

However, behind that serial calm, there is a turbulent foaming ocean at the quantum scale supporting wavefunctions of potential, unfettered by restrictive conditioning of linear time. These structureless, amorphous abstracts lie quite beyond our conscious awareness. Although influenced at individual microscopic levels, the cumulative results over many outcomes become a meaningful trend. Emergent states, leveraged by complex system effects, tend toward significant deviations from what might otherwise have been random conditions. Nevertheless, they are available for selective sampling by Sentience — shaping, sequencing, and manifesting such potential to the benefit of sentience.

We are nonconscious participants influencing the necessary fine-tuning and detailed retro-creation of the historical shape of the working universe toward its Now condition. We are supporting a past that leads to the manifestation of the present as it is for us now. So, even though historical events may not appear to have present-day causes affecting them, in such cases they already did!

One might question how the thought from an unmanifest consciousness could affect something that is "real"; yet in our humble way, we are doing that all the time. Our mind conjures up a mental image or concept of what we want to happen, and we cause it, or not. Our body-mind is a continuous interface between conceptual creations of our mind and the physical manifestation of their reality through our bodily interventions. In our medium-scale way, we are constantly manifesting thoughts, inspirations, concepts, and yes, dreams. At the beginning of our day, we wilfully don clothes and project ourselves into our environment from within those clothes. Likewise, in this virtual existence of ours, we, as Sentience and spirit, don a body for the duration of life, which is manifested from microscopic parts of the reality we enter, and which allows us to project ourselves into, experience, and interact with that physical reality. We dream the dream, create the creation, and perform within it.

Looking into the process of creation, we cannot fully understand it because we are a component of that process, which extends beyond us; our linear knowledge, both objective and subjective, is not sufficiently profound or mature. Nevertheless, we endeavour to understand existence by evolving through the various perspectives of superstition, metaphor, philosophy, and now, science. In this account, we embrace the opportunity to step away from our restrictive dependence on the paradigms of time, locality, duality, and separateness. To model how existence might work, we look at analogous processes in the natural physics of life. These processes, stemming from small, simple beginnings, can cause major creations and shifts in their reality, and appear to emulate the subjectively perceived manifestation and operation of what we experience. We can acknowledge that the deep thinkers throughout the short history of our society may already have subjectively sensed similar understandings of the creative process of existence. The difficulty has always been to render those individual first-person raw understandings into a rational, conceptual form that can then effectively transfer into a contemporary, linear-thinking, puzzle-solving consciousness; and thence, to convert them into a suitable language for sharing among other sentient beings for consideration, debate, and further processing.

Perhaps only in this age, and beyond, are concepts of the workings of physical matter sufficiently progressed and distributed throughout humanity, that reasonable science-influenced, rational-based analogies and metaphors can be conjured. These informed ideas can more effectively replace the resplendent, poetically embellished, and superstitious metaphors of old, which historically – and often literally – are still used to interpret and disseminate the intuitive, raw understandings of that age of how the universe works. Only now and onward can we expect to get closer to capturing a more adequate, objective, and insightful conception of the complexity of the process we call reality. We can appreciate the role that conscious awareness and enquiry have in the business of manifesting form and the analogous mechanisms by which we can comprehend such processes. We are learning that the essential tools of creation, and indeed of all existence, are consciousness, entanglement, and change; and that the fundamental agency mediating the process is our nonconscious selves in conjunction with the universally distributed network of Sentience.

We are considering the prospect that multitudes of unmanifest versions of potential reality exist. From these, one virtual reality is sampled for conscious experiencing. The act of sampling and manifesting an existence from this ocean of potential promotes the elements of that reality to form entangled interrelationships, which influence the entire networked structural and historical temporal fabric of the bulk of our perceived reality. That action of sequential selection takes place anywhere, anytime, and always. Thus, every participating sentient being contributes nonconsciously to the distillation and crystallization of the perceived collage depicting the continuum of reality being experienced.

If something needs to take place in our past in support of the present, then for us, it has already done so. The present is just taking place, though not everywhere at once. Our conscious mind is permanently living in a past version of reality that is already a half-second old; if we bask in the sunlight, its light emitted several minutes ago; if we look at the stars, the light we see now was emitted anywhere from tens to billions of years ago. The universe itself cannot react to any event faster than lightspeed. We have no conscious ability to sense a non-relative present, an absolute Now in real-time. The present is relative to the locality, the sentient, and the sense. Indeed, as we discovered, it is so relative that a window of Now cannot even be defined. As with infinity, the concept of Now is a handy conceptual tool that we use but does not exist in any objective manner. We only have our out-of-date sensory experience, an intractable history, and an indeterminable future.

By being aware, by experiencing the operation of existence within the universe, and by inventing solutions to the "how-does-that-work" questions that we ask, we eventually discover the answers. Those answers must be consistent with other truths. Still, in those valid discoveries, we are also permitting those answers to manifest to the same relative sufficiency-of-detail as that determined by the discovery. That creation process continues, despite our being oblivious to the nonconscious connection between our mind and the universal Sentience. We are mostly not even aware that part of our mind is indirectly, and cooperatively, inducing the appropriate level of detail for the present and the past of our experiential existence. Unwittingly, we continually co-create our reality of Now, and the sum of all its histories.

As one final point, for conceptual simplicity, we originally considered Sentience as a single universal field of unconditioned consciousness. Now we can recognize the possibility that Sentience is not some overarching entity that could be mistaken for a deity. Instead, it represents a ubiquitous, multidimensional web — a neural network of interwoven communication strands entangling the nonconscious of each sentient being. Remarkably, by way of this universally distributed sentient network where every node represents the nonconsciousness of an individual sentient being, we cause ourselves.

We have been discussing the principles of how the universe might work — its operational functionality. Next, we will look at why the universe works this way.

PURPOSE FOR BEING

By now, we recognize that our world only seems real to us because we, and all our means of sensing it, are made of the same stuff. At the leading edges of theoretical and experimental physics, the basis of all our reality – of every one of the individual atoms making up the entire universe and each of us – appears to comprise nothing but vibrating swirls of probabilistically influenced energy. Moreover, the various conventionally accepted barriers that seem to limit or contain our existence may not be as insurmountable, as impenetrable, or as perfect as were once assumed. Such now questionable boundaries include:

- The speed of light
- The three conventional dimensions of space
- The arrow of time
- The temperature of absolute zero
- The total vacuum of space

It seems that the development and stable operation of our material universe may require various nonlocal activities to be continually taking place behind the backdrop of spacetime, and beyond those conventionally accepted physical limitations. Such activities then impinge on our known physics in subtle, indirect, and currently inexplicable ways.

The realization is dawning that the physical existence of the universe and all it contains is not some random event but may depend on a sampling process of sentient entanglement at the quantum micro-scale level, extracting reality from the superposition states of innumerable possibilities. This process requires continual change, and change is information. Thus, the basis of our existence depends upon the extraction, generation, and projection of information. Consciousness is the information processor, experiencer, and enhancer. Sentient beings indirectly participate in selecting, preserving, and sequentially ordering that enhanced information.

We have noted that by the process of meditation and inward contemplation, we could accommodate subjective, non-dual awareness and glimpses of the supportive existence of an unconditioned, enabling, universally distributed field of consciousness that is currently beyond our ken. Furthermore, it seems possible to sporadically tap into this limitless field of knowing, identified here as an extradimensional field of Sentience, which usually is not consciously accessible, and which seems alien to our existential physical experience. Although subjective, but drawing on modern scientific principles, this recognition of the process of existence has parallels with the many ancient themes of a spiritual and metaphysical nature, handed down throughout the generations. These are similar to the themes that once supported the now-obscured foundations of most spiritual traditions, both theistic and not.

Furthermore, we are coming to realize that such a Sentience-inspired physical existence as we are experiencing is not a cut-and-dried, ready-made, completed world made available for our enjoyment and discovery. Instead, it is a developing thought, a lucid dream, a creation nurtured within Sentience, which continues its evolution by requiring the co-creative participation of its sentient members, which include us. Our participation is an essential contribution to the creation and development of the physical working reality that we think we are merely discovering. Each one of our nonlocal nonconscious minds is an integrated, full-membership, holographic-like projection of that universally distributed Sentience. We unknowingly participate in co-creating our physical existence with, and as, executive agents of this Sentience.

Moreover, it seems likely that the collective plurality of the nonlocal part of every sentient unconsciousness – the nonconscious realms of psyche and spirit – forms the distributed, entangled, and metaphysical network comprising the web that is Sentience. Thus, along with evolving our reality, we, as foundational facets of Sentience, are mutually evolving ourselves.

Our Purpose

Recognizing this new perspective, we can now improve our fundamental understanding of our being. We are poised to address that age-old enquiry, "who am i, and what is my purpose?"

Traditional religious wisdom may suggest our best purpose in life is as children participating in playground activities, to enjoy Being and avoid suffering; though some of the more draconian religions might disagree. Regardless, most of them would undoubtedly mandate god worship as a significant activity.

However, most humans anecdotally report that, in retrospect, the richer a physical and mental life is in a variety of experiences, including both ups and downs, the more rewarding is the being. In the Buddhist tradition, pain and suffering mainly result from grasping for, and ignorance of, the insubstantial and impermanent nature of existence, as exemplified by materialistic and dualistic thinking. If that ignorance were replaced by a deeper understanding of the true sentience-dependent origin and the impermanent nature of beingness — and by extension, of us — then a journey of wonder, joy, and fulfillment would accompany such enlightenment. That journey, learning, and transcendence is our real goal. To rise above ignorance, to transcend irrationality, and to discover ourselves for what we are — an essential part of the process of carving out a self-fulfilling creation — a dream incarnate.

Particularly in religious contexts, a person is sometimes described — or self-described if of evangelical bent — as an instrument of God. Regardless of such narrow religious connotations, the term "instrument" is an appropriately secular one to use here for a sentient being, in that an instrument can operate bidirectionally. It can observe and perhaps record or transmit findings concerning the nature of its focus; it can process and relay its experience, as in a microscope, allowing observation of the small but close, or as in a telescope allowing a glimpse of the large but distant. An instrument can also physically intervene or modify the object of that focus or its situation, as with a probe, scalpel, or wrench. In the sense of a smart tool for intervention, we can be both. Thus, we can appreciate that in a fuller context, sentient beings are instrumental to their existence.

In the context of this world, all human beings — not just a select few — are instruments experiencing, probing, and interacting with our environment, our reality. As an instrument, our consciousness - or more specifically our unconscious — acts as a bidirectional translator. It serves as an interface between the phenomenal and the ethereal. In one direction, physically related information from our senses and sensors is perceived and processed to form mental images, maps, and meaning — thought. In the other direction, thoughts and existential awareness provide mental motivations to manipulate our environment physically. Our nonconscious is fundamental to this two-way exchange.

Furthermore, not only humans, but all lifeforms, all sentience, and all consciousnesses, perform to some degree, as instruments of Sentience, and the web of Sentience acts as the instrument of all consciousness. We are experiencing what is and creating what needs to be to continue and evolve. We experience being the observer and the observed, dancer and the dance, artist and the art. By enabling Sentience to be the artist that paints itself into the picture that is our reality, we become the ultimate existential self-portrait.

Creativity

A materialist or physicalist view of the purpose for being is that our existence – the existence of life at all – has absolutely no meaning and is just a curious accident of a stressed self-organizing mass-energy system that accidentally appeared and improbably evolved. However, such a nihilistic view misses the point. A dream is the self-expression of the unconscious mind, thought the self-expression of the conscious mind, and inspiration the self-expression of both. The purpose of creation, existence, and life is the furthering and evolution of sentient self-expression. Reality then is the self-expression of all consciousnesses, and these intelligent energies comprise Sentience. These energies are not the sorts with which we are familiar – those that generate heat, light, and force. Conscious energy is a subtle, dynamic, interacting intelligence, which overarches the material, and currently appears beyond the capacity of our comprehension.

Conscious energy is what got us here, though; to be able to perceive, contemplate our origins, be aware, curious, and ask hard questions. Part of the challenge of expressing ourselves is to try to divine our sentience. Sentience has manifested the physics supporting a universe in which the conditions for life have evolved. Life has formed and birthed a plurality of individually manifest consciousnesses with the intelligence to experience, process, and express information. In essence, they comprise focus points of psyche distributed among individual embodied minds, which act as hidden representatives of the universal web of Sentience.

It is not as though some mischievous deistic entity deliberately deceived us into thinking that the physics of our reality is complete, merely awaiting our discovery. In our ignorance, we deluded ourselves into thinking thus – that matter is solid and existence real and pre-ordained. We, not some external agency, are the ones focused on dualistic, materialistic perceptions and paradigms as descriptions of our reality; that something exists or does not, is either good or bad, or must be one way or the other. What we do not understand, what we cannot fully comprehend, is that our role in life is not the discovery of what is – though it may appear that way. Led by awareness and curiosity, that role is the creation of reality by determining a causal essence for what we perceive to be already before us. In this effort of understanding and discovery, in that coherent, focused authentic awareness lies the creation of the formative physical detail of our reality. What we interpret as discovery is in effect unconscious creation.

We have looked at creation and its relationship to art, as a metaphor explaining the nature of some of the fundamental aspects of existence in spacetime. We are regarding creativity as being a natural, self-actualizing function of consciousness, and, along with survival, a motivating force for all sentience. We mentioned curiosity as an attribute related to discovery – it is effectively fundamental to the urge to learn. Curiosity and creativity are intricately linked, and by adding discovery into that mix, one sees a triptych of related sentient attributes. These encapsulate what we have been talking about here, the behind-the-scenes mechanism responsible for existence, as we know it. Sentience observes, and, curiosity aroused, extends its new awareness to discovering what its observation entails, how it might be caused. In defining the causal mechanism, sentience permits the manifest creation of the observed.

Creativity, the desire to create, like curiosity, is a fundamental and sacred part of sentient consciousness. At an individual level, it may not always be the highest priority, though, particularly if the sentient consciousness is otherwise occupied with the priorities of physical life, such as survival. Humanity took several million years of adaptive evolution, a much more recently developed neocortex, and tool-making skills to help manage their survival needs. The desire to create blossomed as soon as the basic needs of survival were generally met, and humankind had time to devote themselves to such higher occupations. Only some fifty thousand years ago might that have started, when the urge to wonder and create shareable ideas and art was being cultivated. That ability had first to evolve, to be enabled, and then developed, although it arose out of the prioritized innate desire to survive, discover, communicate, and advance – self-actualization.

Self-actualization

In the last century, Abraham Maslow theorized a hierarchy, a pyramid of human needs to depict the route to self-actualization, starting with the priority of broad base-level survival needs. Each subsequent stage had to be satisfied before moving to the next. Right at the top of this hierarchy – highest tier though relatively low priority – were the needs for spiritual understanding and self-actualization, and it is at this level that creativity for its own sake would fit. Creation seems a compulsory product of sentient awareness – of consciousness – and likely the primary motivation for our universal Sentience.

The purpose then, of the thought that underwrites our existence – this lucid dream – is self-actualization through creative expression. As individual components of a sentient society, our purpose within that means of expression is to work with the detail, to experience what could sustain the particulars of the present manifest condition, and collectively feed that information back to Sentience. It is also to augment, fine-tune, and guide the evolution of the universe concept toward enhancing its virtual form. It is to enable sufficient detail to support and evolve the thought, and to enable its robust manifestation. Our roles and responsibilities are to compassionately experience, enquire, discover, self-actualize, and be our best in life. From our partisan perspective, it would seem better to do this from a joyous perspective, though this may not always happen – joy is not a birthright.

When you get a heavy cold or worse, the experience is not pleasant. Extracting joy from such illness is difficult. Yet, from the invading virus's perspective, if it were capable of such emotions, it would probably be joyful with self-fulfilling expression. It has been able to run rampant throughout your body doing what it does best, co-opting resident cells, successfully reproducing, and likely being entirely unaware of the collateral damage to the host's condition. From the broad viewpoint of nature, there is no dualistic sense of absolute right or wrong. The issue of good or bad becomes one of what suits or does not suit the perspectives of individuals, groups, or society making that judgment and how each one of us creatively responds to change, challenge, and opportunity.

Therefore, that our purpose is to live in joy is more of a partisan hope than an entitlement. Nature may not share such partial priorities of individual perspectives and can be much more equitable. Life can dish out opportunities for suffering with as much equanimity as opportunities for joy. A Buddhist proposition is that life is already full of suffering, though it is mainly due to ignorance and illusion within the afflicted individual. The key mitigator to that suffering is awareness;

how each plays the hand dealt, bad hand or good; how the individual can find ways of self-fulfilling even while experiencing suffering; and if that individual can become aware of, overcome or cope with the underlying reasons for that suffering. Even unfortunate individuals who seem unable to find a way beyond suffering are still fulfilling the role of experiencing, and perhaps learning and evolving in the process. They provide continuous feedback to Sentience and are fully engaging in a collateral part of the creation and evolution process.

Evolution

An opportunity that most often emerges from the stress of suffering is to change – to evolve to another stage, one less adversely impacted by whatever was the cause of that suffering. This sort of stress management is one of the main driving forces behind nature's biological evolution. If its environment were to stress an organism critically, it would move away if it could, evolve into a more capable form appropriate for that changing environment, or die; a process the philosopher, Friedrich Nietzsche, summed up in the axiom "that which does not kill us, makes us stronger." Likewise, the stress of psychological suffering, if we were to remain ignorant of the cause, may induce us to run from the circumstances, to try to forget, or to flounder, letting the stress exacerbate the suffering, pull us down, undermine our well-being, and shorten our lives.

On the other hand, becoming aware of the source of stress and suffering may enable us to take on the responsibility to evolve mentally and spiritually into a more expanded, less burdened mindset. Some stress is a normal and often a motivational part of life, however, many instances of stress turn out to be self-imposed and essentially self-remedial, resulting from patterns of unrealistic, egoistic, inviable expectations and personas.

The main characteristics of the driving force behind Being, point to expressions of striving toward higher purposes. The way to approach these higher purposes is through informed evolution. Thus, the prime objective of Being is self-evolution – self-development. A characteristic of consciousness, beyond survival and self-awareness, is to experience and evolve; to grow, mature, improve, and create. In doing that, the sum of its information content increases – an inverse relationship to entropy. The process of life is all about swimming upstream, against those entropic currents. Creation and change must be part of that process, otherwise, there is no new place, form, or potential, toward which we can evolve. The universe is part of us and us of it, so it too must evolve. The natural evolution of consciousness contains the seeds of compassionate creation as part of its self-evolvement.

Despite the woodcarver metaphor frequently used here, we should not visualize the creative evolutionary process as an individual initiator creating a separate object. The process comprises a mutual, relational involvement of constituents that collectively form Sentience, of which we are always already a part. There is no separate object or being, only our collective nonconscious network forming Sentience. Within this Sentience is all that there is. Furthermore, at the psychic level, we individual sentient beings are the metaphoric fingers and tools of that collective source of creative intelligence. In effect, we are Sentience, nonconsciously working on the details of our creation, which includes ourselves. Our mutual purpose is to evolve, and we exist because we choose to.

Performance

In the modest context of participating in existence – the creation and evolution of the universe – one might question if we sentient beings, as instruments of that ongoing process, are fit for participating in the overall purpose. Let us once again remind ourselves that all manner of sentient beings, all consciousnesses, act as agents of the creative process, not just the human ones. In terms of all such sentient beings, including those we are not aware of, it seems as likely as not that we must be fit for that purpose since our evolving universe exists, and we within it. As far as we know, the whole system, though highly improbable, seems spectacularly effective.

Sometimes one might wonder if we humans are a net contributor to or detractor from the creative efforts of the universal gene pool of all sentient beings. Few would have the hubris to claim that Homo sapiens must be the best in class in our mental and physical functionality and our contribution to existence. We are what we are and find ourselves in a state of philosophical ambivalence sprawling across and vacillating between the interior and the exterior self – the Being and the Doing. A definite answer is beyond our reach, though the challenges that we, as elements of Sentience, have set ourselves for all sentient beings to meet should be within our means and responsibility to resolve. If we, as Sentience, cannot create, experience, and evolve a manifest universe and life force, in part through human contributions, then perhaps we can do it through other forms of sentience better suited to this purpose. As Sentience, we are unlikely to fail, this being our dream after all.

Possibly we humans are still too young and immature a species to be assessed in this matter, having been around physically for little more than a million years. Our most recent significant brain and mental evolution is not yet two-hundred millennia-old, and our culture less than fifty. That is a blink of an eye compared to life's gross evolutionary history, measured across several billion years here on Earth. The U.S. Population Reference Bureau estimates that over 100 billion of us have lived on this planet since humans evolved, with more than seven billion of us existing now. From our limited and admittedly human-biased perspective, we have no comparable reference. We can only surmise about the effectiveness of this squabbling, compassionate, bumbling, intelligent, self-absorbed, pluralistic, pre-rational, trans-rational, label-ridden, linear-thinking, materialistic, dualistic, and immature sentient species that is humanity. What a glorious paradox we are! An enigma poised, evolutionarily speaking, on the brink of disaster – much of our own making – and salvation. We are teetering on the moral edge between self-indulgence and self-awareness, between materialistic ignorance and spiritual awakening. It seems for every "bad" characteristic we have; one can think of a corresponding "good" feature; for every negative, a positive. That tension between polarities, that stress, is what stimulates evolution – evolve or die.

We need to balance population growth and consumption ethically with all other legitimate claimants from sentient life forms and environmental infrastructure to planetary and eventually galactic resources. We may not have a good record as stewards to date; however, we are becoming more aware and slowly improving our stewardship. Excepting an extinction event, the main challenge for our species in the future is to survive to evolve and evolve to survive, to encourage physical, mental, and psycho-spiritual transcendence, and to contribute to the grand design of our reality. Whether we are successful or not in that endeavour, the universe will continue with or without us for as long as there is any form of sentience to witness.

Maturation

Evolution generally describes a process of incrementing toward a more advanced stage, whereas transcendence stresses surpassing previous limits of experience. Although of similar meaning, evolution may apply to a more biophysical or environmentally driven process, whereas transcendence points to a more metaphysical or psycho-spiritual process – one driven more by mental strivings. A key point for both terms is that they describe an advancement process that does not abandon previous stages. It is not a question of jumping from one state of body or mind to the next, leaving the first behind. The new state emerges from and includes the previous one; indeed, it is founded upon it. Each state remains present and needs appropriate incorporation to form the new integrated whole.

Advancement proceeds in something like a dance sequence – experience, differentiate, integrate, and repeat – whether the process is physical or psychological. It can occur as a maturation process such as in consolidating and fine-tuning an already stable phase, or as a spontaneous creative evolutionary jump, inspired by capricious, inspirational, or stress-related factors. Humanity currently appears to be, physiologically and functionally, substantially into the maturing evolutionary process. However, psychologically and spiritually, we can anticipate further evolutionary jumps. After the last several millennia, we may now be at the point of emerging from one such stage and heading toward another mass-transcendence of consciousness – even within this century.

As we have observed, our species is immature, often seeming at the level of a pre-schooler still embroiled in egoistic tantrums, conflict, and magical, superstitious thinking. Physically, we have become quite adept at taking things apart and creating ever-more elaborate technologies, although generally, our mindset remains egoistic, and our philosophy materialistic. Frequently, many of us appear to be barely out of the self-indulgent tantrums of the terrible two's. As a species, we may still have a long way to go before claiming adulthood – and even further, wisdom.

Like the stages of emotional maturity development, we discussed earlier, any large population will exhibit variations in wisdom around the society's median values. These variations might follow a classic statistical distribution, such as the simple bell-shaped curve. Imagine if such a curve were to describe humankind's transcendent wisdom or psycho-spiritual maturity. The lagging edge of it would include a portion of our species that are still at a pre-rational level of mentality, working full-time on real or imagined fight-or-flight survival strategies. A smaller, more enlightened portion of the population might represent the leading edge of the curve. They can authentically think holistically, altruistically, and compassionately from a global perspective. Their elevated perspective would integrate not just humanity, but other life forms, the environment, this planet, the potential of the entire universe and perhaps beyond that; all that there might be.

During the ongoing evolutionary process, the median – that significant central body of the bell-curve, the main bulk of our species – may aspire to move toward the position of the leading edge. If there were such a thing as a fixed goal of full enlightenment or psycho-spiritual evolution, then gradually, the leading edge would get there, followed later by the median, and much later by the tail end of the population.

However, it is not as simple as that. Neither evolution nor transcendence is a race to a fixed goal. There is no final evolutionary destination, assuming the survival of the species at all. What matters is the journey; the path is all. Over time, this imaginary population bell-curve may undulate like an inchworm, onward in the positive evolutionary and transcendental direction, with the leading edge first, followed by the median, and finally by the stragglers at the trailing edge. The curve shape may change during this process; it may skew towards one edge or the other. The movement "forward" of the entire curve can vary its pace and may even reverse if devolving into darker ages. The curve may flatten out if substantial differences were to develop between the maturity states of the leading and the lagging edges.

It may sharpen up if some form of unifying evolvement across the entire population were to arise where differences in evolutionary states between the leaders and the laggards became minimal. Fostering that kind of circumstance might have been, still could be, a worthy objective for organized religion.

Statistically speaking, that imaginary bell-shaped curve could be a plot of what the journey of an evolving and transcending species would look like if you had the data. Any individual may be at one extreme or the other of the population bell-curve, and might even lie outside it, though most, by definition, will be within the middle part of the curve – the median. Individuals lend shape to the curve; they form the evolving leadership. If there are enough similarities of psychological perspective among them, they can change the shape and position of the curve along the imaginary evolutionary axis. The individuals in the middle of the curve, by their self-awakening efforts, can shift the bulk forward, toward evolutionary and transcendental development. That is where we are heading as a species.

Sometimes the pace may seem frighteningly feeble and faltering, with devolution right around the corner. Nevertheless, as in heavy traffic, messy bunching and spurting is a natural way forward for a self-organizing system evolving under duress. Evolution is not a smooth process, depending as it does on trigger events: adaption, mutation, and natural selection. Transcendence also tends to occur in spurts, often resulting from trauma, stress, or inspired insight – at both the personal and group levels. Though we may despair at our society's failings, our wars, our moral and materialistic disgraces, we can still take some comfort from our achievements. Many of us are now questioning the old paradigms, the old-school ways of thinking, and are participating as best we can in a movement to embrace higher values. Ultimately, only by being at peace in every member's heart, can society hope for peace in all its doings.

Advancement

That maturation bell-curve may change shape, though it will never go away. As humans, we are not all uniform, and none of us is perfect. Given the natural messiness of life, we might be able to get some feeling for where we are on an undefined evolutionary or transcendental scale, by considering how disparate the evolutionary extremes of our population are. Based on that measure, we have much work to do. We need authentic leaders, not power-mongers. Those leaders themselves need to evolve, to transcend. They need to encourage, assist, and enable the rest of the population. We need a compassionate understanding of the laggards because any large population curve will always have some.

By now, it should be clear that forward steps on this imaginary transcendent bell-curve do not correlate with the state of any religious belief, power, wealth, language, tribe, or nation. Among some of these categories, there may even be a negative correlation. For example, anecdotal evidence over the ages suggests that most of the authentic psycho-spiritually evolved individuals develop from among the less financially wealthy, although some great historical prophets and sages, the Buddha among them, are said to have originated from wealthy families, and only came into awareness after renouncing their material wealth and position. Perhaps that was part of the process. The socially and financially powerful do not seem particularly well represented among such authentic leaders. Continual grasping toward political achievement, material power, status, and wealth being detrimental to profound, humble introspection and compassion.

Advancement through evolution and transcendence was likened earlier to a dance. For a while, some activity may take place though with little forward movement, until suddenly the parties step out with authority and several paces of real progress happen, then the process cycle repeats, though perhaps with variations or errors. Evolvement and transcendence cannot occur in isolated chunks. Like the whole dance, they require and incorporate all previous steps, even the variations and survivable mistakes. The entire holistic history is what makes the dance. It does not disengage from its past performance at the final step, leaving all other steps behind. The whole performance, the entire cumulative historical experience, is retained though transcended – previous steps, missteps, all former states become a foundational part of the next evolving, transcending condition. Eventually, the realization dawns that the dancers and the dance are a single performance, and the sacred purpose of sentient beings is to participate in the co-creation of the universe and co-evolve within that process.

RESOLUTION

As a youngster, i enjoyed "The Hitchhiker's Guide to the Galaxy" by Douglas Adams. Those familiar with this tongue-in-cheek science fiction story may remember that the protagonist, Arthur Dent – the last surviving human being – hitched a ride off Earth moments before it was destroyed to make way for an interstellar bypass. He then found himself interacting with a group of pan-dimensional beings determined to learn the answer to the Ultimate Question of Life, the Universe and Everything. They built a supercomputer – Deep Thought – especially for this purpose. It took some 7½ million earth-years to compute the answer, which turned out to be…42!

I enjoyed the drollness of these tales, which stuck in my young mind as frustratingly humorous. One dissonant feature is, of course, the simplicity of the answer after the arduous trials of Arthur and such an enormously long computation. Another shock is that having produced an answer finally, the details of the original Ultimate Question turn out to have been forgotten. Once i had accepted the dissonance of the answer, i remember being quite disappointed in the choice of 42 as the solution, for it is a very ordinary number. For me, it had no romance; but of course, that was part of the deadpan humour of the whole thing.

Some amusing parallels can be noted between this story and my quest, which initially focused on God, though later migrated to the nature of existence. The idea of a god, as the answer to the mystery of life, might be anachronistic, and the institutionalized need for a deistic concept seems to be more like the answer to a forgotten question, as such was 42. Wrapped in the unquestioned cloak of mystical spirituality assumed by religions is ultimately a misunderstanding of the analogies and metaphors used by the early seekers and elaborators. One may imagine that those original prophets, sages, and gurus – the originators of most religious and secular traditions – tried to communicate their insights and realizations using emotive metaphors of the time. The more authentic prophets may have preferred to instill in their followers the curiosity and self-confidence to each pursue their own understandings and revelations about the nature of the enabling forces that permit and support our existence. Having themselves achieved an abnormal level of enlightenment, and realizing the pitfalls of dogmatic prescriptions, they may have deliberately avoided documenting their findings.

However, that did not work. None of the founders of the great traditions, including the Buddha, prevented the posthumous dogma-publication trap. Some might have resorted to preaching prescriptive commands due to exasperation over the slower of their disciples. However, it was more likely the disciples themselves who resorted to written prescriptions to spread their version of the sage's words, years or even centuries after their prophet died. Though the Buddhist tradition is not immune to dogma, the absence of prescription would have been the Buddha's desire, as is evidenced in the "Believe nothing, O monks" quotation noted in Closing Quotations herein.

Unfortunately, throughout the ages, the concept of deities as being the solution to the question of our existence has been anthropomorphized and institutionalized. The original motivation of the teachings likely intended by many of those originating prophets has been lost. Instead, religion's various prescribed forms of autocratic dogma, superstitious ritual, and spiritual materialism, mostly destined for careless consumption by many, have replaced them. Much of religion's current fare provides an active venue for social interaction, a degree of comfort through ritual and platitude, and generally, a positive balance of moral direction. However, it seems to contribute little profound guidance for evolving the innate psycho-spiritual comprehension of the individual layperson or indeed the species. Much of the original morality and socio-ethical standards appropriated by religions could still be supportive of an immature though evolving society. Unfortunately, a history of corruption, immorality, and infliction of barbaric violence in the name of their god, has tended to subvert such benefits. Nevertheless, even now there appear to be some individuals within such traditions who are trying to reset their chosen religion's direction to encourage such support.

However, traditional religious beliefs are not the only route available for nurturing a society's morality, and one might expect an evolving secular society to develop such positive mores anyway, as part of its survival and natural transcending process. After all, virtues such as compassion, love, and wisdom did not arise from religion; religion arose from the followers of a few enlightened individuals expressing those virtues. For most religions, such mores then became prescribed, acquiring archaic authoritarian abstractions and superstitious encumbrances in the process. While that may have been appropriate several thousand years, they no longer seem an effective vehicle for encouraging contemporary humanity's authentic psycho-spiritual advancement. The historical result of this misstep is that the psycho-spiritual evolution of our species, which should be advancing together with our material abilities, remains stagnant, fossilized in the dust of ages. The subsequent evolutionary imbalance in awareness threatens the welfare of our spiritual and psychical consciousness, and the resulting ignorance generates unnecessary distress and suffering among the world's population.

On the positive side, there does seem to be a palpable upsurge of interest among many individuals in modern times, a tangible shift in consciousness toward authentic psycho-spiritual matters such as compassion and inclusion – not so much as religious fare, but more as a profound philosophical and ethical matter. There appears to be more evidence of increasing pockets of truly mind-expanding liberality in the widespread search for more secular psycho-spiritual comprehensions. Individuals and groups, both outside and within religious organizations, are following personal paths, making fresh tracks towards increasing self-sourced enlightenment with associated enhancements in self-confidence, self-awareness, and self-actualization.

Certain phenomena are not explainable as arising from physical origins. As noted earlier, such awkward subjects include consciousness, thought, memory, mind, instinct, inspiration, intuition, and the like; however, they can be perused in contemplative meditation. Conventional scientific materialism cannot yet go to such awkward places as these, and so their explanation becomes characterized by many mainstream scientific minds as unresolvable or irrelevant issues. Conversely, there are numerous areas where inspiration and contemplative enquiry are unlikely tools to yield clear, detailed, and repeatable technical answers. However, on occasion, there does appear to be some rapport between authentic understandings gained through the protocols of objective physical science and those profound perceptions gleaned through subjective contemplative enquiry.

That there are convergences and divergences in the application of objective or subjective methods of enquiry is not the issue. The issue is the mostly unquestioned and exclusive dominance – habituated into the recent minds of humanity – of the one protocol over the other in those areas where both may have the potential for synergistic mutual contribution. This imbalance leads to polarizing extremes of either a materialistic- or a superstition-dominated dualistic sense of reality, one where individuals might perceive difficulties of comprehension as looming so large, they feel they can only wait to be told by others what to think – and, absent adequate education, this condition impacts matters both scientific and spiritual. That polarized psychological condition, occupied by many of us on this earth, needs to change. At least it needs illuminating, so that individuals with enquiring minds may find more support for the rational pursuit, both objectively and subjectively, of their personal quest. Consequently, they may find a place of more balanced truth and awareness – a personal coefficient of existence.

The appearance of and tolerance for another Messiah seems unlikely in the present time. Perhaps, though, enough self-illuminated people will become humble beacons engendering a significant positive psycho-spiritual social disruption, a massive transcendence of consciousness in our society's asymmetric evolutionary progress.

Meanwhile, there is truth of existence that can be found by each one of us. It is not an absolute answer to the question of life, not a one-size-fits-all kind of truth amenable to recording and prescribing. It is personal, subjective wisdom, which, while having generalized veracity, remains specific to each individual. Such truth can initially reveal itself through personal experience, contemplative enquiry, inspiration, and intuition. What discoverers of this sudden truth may find is that their glimpse of Sentience-engaged existence brings about a personal perspective of clarity and certainty. Such a view points toward what a few have been attempting to explain throughout history; an awakening that can never be equalled by blind declarations of belief in that which is not their own.

Furthermore, the answer is not 42!

The God Question

That brings us back a full circle to the beginning of this quest, to my original God question. This question, prompted by a profound self-manifesting poem that originated overnight in my sleep, is what propelled me seriously into this spiritual quest. The quest initially drew me to an introductory attendance at a meditation class where the meditation leader had surprised me by asking my prime reason for being there. The impromptu answer i gave then – albeit naive and prejudiced by previous experiences of an imposed God – was that i wanted to understand how God made the universe, and who He was. Answering this God question seemed to be a tall order at the time, although i was sincere. Over the ensuing years of meditative practice and contemplation, i have come to realize i have answers that are satisfactory to me. However, they have proven to be much more secular and profound than i had anticipated. I probably always had them, hidden deep down, as each of us does, though now i recognize them; they have become realized.

There is a phrase "to beg the question," which describes a question framed to include assumptions about its content that prejudice the answer. A classic example of this, thought to be a construct of Jesuit monks, asks "how many angels can dance on the head of a pin?" It contains the presumptions that there are angels, they can dance, and they occupy some finite dimension. Similarly, my initial God question contained the inherent presumptions that God existed as a separate entity, was person-like, and is a He. What i discovered is that He did not, was not, and is not.

The deistic concept can be misleading on so many levels, though we cannot blame a god for that. After all, humankind provided the names and characters representing those assumed mysterious, controlling, omniscient presences. Distant deities or a single omnipotent deity appeared necessary to explain the world, the mystery of our existence, and to keep good order among the population. This successful concept was augmented by those humans wishing to exert more control over the social behaviour of others through threats of wrath from such an ultimate authority figure. The term god may have represented initially, in whatever language of the time, a desire to label and anthropomorphize the presumed incomprehensible activating force or energy behind our being – the principal cause of the birth and the ongoing existence of the universe. Over the centuries, though, that mystical label became open to considerable re-interpretation and abuse by diverse religious politics and entwined with superstition, ritual, and dogma.

We can look past this deistic label and examine the entity to which it attaches. Even the term "entity" is prejudicial, a word reflecting materialism in the same sense as wisps of energy are still called particles. The more secular alternative, Sentience, was introduced here to represent a scalar-like field of universal consciousness. It comprises an all-permeating, all-encompassing, ubiquitous ground of being and an inconceivable extradimensional field within which the universe is nourished. It is, we are, at once stage, play, actors, and audience – the performers and performance. We are an integral part of Sentience, which is where our reality is conjured. Sentience is in us, and that is why we savour the experience of existence. However, i acknowledge that the term Sentience is just another label that can mislead and limit like any other.

One might wonder what a god's view of us might be, observing our ant-like efforts to exist in this vast place we call the universe. Many of us believe we have little or no connection to the rest of the universe, and we uphold a subservient, distant, worshipful, fearful, superstitious relationship to the kind of god that some still accept. Such a god might see us as a group of struggling, egotistical, bickering delinquents, hemorrhaging our true power, our future, and our blessings into illusion, into a meaningless materialistic mirage. Surrounded by the magnificence of this manifesting universe, we might be squandering our birthright on irrelevancies. Some perceive this version of life as so unrewarding, so intrinsically unsatisfactory, that they try to escape from it in the various ways offered by despair, denial, and depression. These are the victims of an imposed or self-induced illusory perspective of the exterior.

However, the real perspective is not the external materialistic one. It is only accessible through the interior. Attempting to see the universe and ourselves from a god-like perspective is to see our own hands. That perspective would be as natural as looking at ourselves, for we are the seer and the seen, the witness and the witnessed, the creator and the created. Any god-like perspective would be from the vantage point of universal consciousness and would contain our experiences, our dreams, our thoughts, ideas, concepts, and visions. We look into the mirror we call reality and see a manifest version of ourselves. Such a god's perspective is thus our introverted view, for she who is we, looks back at us and sees one.

When we look past the material, and experience existence unconditionally, we can feel and witness our oneness with all that there is and can be – firsthand. Only through firsthand inward experience will we comprehend an accurate representation of the perspective credited by many to a creative force or god. It is a perspective of consideration without viewpoint, unconditional without limit. No jealous vengefulness exists here, just the love of an artist for their art, the altruistic giving of oneself to creative expression; to extending, experiencing, improving, perceiving, and to loving one's authentic self. Such a viewpoint might then become recognized as our innocent nonconscious identification with our shared universal Sentience, the spirit of our very being.

> "In the beginning, there was neither existence nor non-existence.
> All this world was unmanifest energy...
> The One breathed without breath, by Its own power,
> Nothing else was there..."

This description of creation reflects many of the post-modern concepts we have been discussing here. However, it comes from the Hymn of Creation in the Rigveda, an ancient Hindu sacred text dating back some three thousand years.

A prevailing paradigm, perpetuated by ancient texts and religious lineage, and assumed by science, is that the universe is already in existence, and we have appeared here like houseguests to live in it; that it is a magnificent mansion in which we find ourselves – merely to explore room by room, and perhaps determine how it came to be. However, we now realize that the reality of how the universe came to be is so much more exciting than that. We – the inclusive All sentient beings – are privileged to be the designers, builders, as well as residents of the said mansion. The universe, spacetime, all components therein, and all sentient beings are borne of Sentience. Asking where Sentience was before the universe began, is like asking where the dream was before becoming aware of it. Though the question seems a natural one, it contains the unstated assumption that Sentience, much like the god concept, must have a location of existence, an abode separated from us. Universal Sentience simply *is* and mandates spacetime and matter. The possible way that might occur has been the subject of these chapters.

In the process of my quest, i came to rephrase the God question as not how He made the Universe, but how the "reality" in which we find ourselves might have become. Still not a trivial enquiry, though now perhaps framed less prejudicially. We have already discussed some broad thoughts on this aspect. They essentially boil down to a system of ongoing creation. To the extent that any part of the virtual universe needs to become existent to support the experience of it, it is Sentience's imaginings and ours, that directionally influence the necessary historical generation and behaviour of the physics of proto matter. It is a reality that becomes manifest to our senses when subject to coherent, informed awareness by sentient beings like us. Until the conscious gaze of such attention directs itself at it, the universe remains a virtual place of broad-brush imaginings and strokes of potential, yet still of awe-inspiring magnificence. The wandering, wondering gaze of our awareness informs Sentience in the transmuting of virtual into sentient-beneficial reality. That is what we do.

Who is God?

Perhaps the most significant part of the "God question" was who is He? Here, the challenge is to look inwards to find our reference and validation. We can question if our moral conduct throughout life would be undermined or enhanced by having a richer understanding, a more profound and intimate relationship with the catalyst of our existence; to peer behind the mask of God.

God is revealed as an ancient label attached to an ill-defined concept. A label, like a graven image, can quickly limit profound understanding, subdue further enquiry, and add distance. Even the term Sentience is just another label, however, by using contemporary concepts, i have tried to remove elements of fear, mysticism, dogma, and superstition from the concept of existence that religious labels tend to represent. While Sentience may still masquerade as the mind of God for some, it represents all that there is and is not. No other reality exists, nothing – literally no thing – save universal consciousness. No limiting label can hope to describe such a limitless feature, and any attempt to circumscribe the content will always leave something outside. Complete comprehension of the concept of a universal network of entangled collective nonconsciousness may not be achievable, but that does not make it deistic.

The greatest miracle and mystery of our world is consciousness – awareness and thought. Sentience is universally distributed unconditioned consciousness, and our world resides in it. It permeates all because it is all, and all is it. Sentience creates, we participate, we create, and we are. Whenever we look inside ourselves – into our interior world, our fundamental cause for being – we find the ubiquitous presence of Sentience contained in us and by us. This "all" is an interwoven, interlocking, encompassing, holistic, and fully entangled energetic universal network. Within it, there is no end, no beginning, no limit, no boundary, and no containment. Conceptually, we cannot define such a picture of existence since the tools of our imagination are both limited and limiting. We find ourselves as small fish trying to depict the ocean; it is everywhere and unbounded.

No wonder those ancient, enlightened sages, or more likely their disciples, attempted to anthropomorphize any experiences of the ethereal by calling that enlightening awareness by many names, some of which would eventually translate into deistic identities. Subsequently, all manner of human-like characteristics became attached to these deific labels, depending on circumstances. Now, we have seen who this god is; it is neither a he nor a she, nor a who, nor an it, but Us. Within each of us is a partial share of Sentience, and the whole includes us and indeed is us. We partly represent it, individual by individual, and collectively in conjunction with all other sentient beings occupying this virtual universe. We are all unconsciously engaged in doing our part to evolve and fulfill the dream of Sentience, turning this virtual existence into manifest, sensory reality.

Who am I?

My God question did not include this enquiry – "who am i?" It should have done though because once we have a handle on who God is or is not, and how and why we exist, we have effectively established the basis for a profound realization of who we are. The question "who am i" is also a familiar mantra, a humble internal questioning often used during meditation practice, thus it would have been entirely pertinent had i asked that. Perhaps this was the unconscious motivation, the unrecognized and unstated reason for my being at the meditation centre that first day. However, i did not think to ask it at the time, so this section represents a gratuitous answer to an unasked question. It might be the one question that people ask themselves the most when reflecting. Perhaps they ask it more frequently than they ask who God is, and though these may appear to be different questions, they have pretty much the same answer.

Our Exterior

Every one of us comprises an interior and an exterior. Convention defines my exterior as the skin surface of my physical body. However, my exterior also includes my interrelated surroundings and interactions that support my existence and involve me with other beings, the country, environment, world, and ultimately the entire universe – limitless. When i think about it more, it also includes my entire timeline from birth to now, and all the causal links coming before and during that existence. Also, in the past, how i have related to others, sentient and not, with my parents, my ancestors and theirs, all history, and all pre-history.

My entire exterior existence is the history of all events leading up to me, as my associated timeline spirals through spacetime. My physical exterior is that spiral fully entangled with the sum-of-all-histories of the universe. Like you, i am physically the sum of all my related experiences, as well as those links preceding and ultimately permitting me to be – that is the entire networked foundation of my existential experience.

That is the exterior me as manifested in physical terms. My physical manifestation requires the extraction of energetic particles from the subspace quantum froth that underlies all reality, and which amalgamate to form quarks, electrons, protons, neutrons, atoms, and the stars. Their exploded dust became the physical content and circumstances of you and me and all our interdependent surroundings. The atoms, molecules, cells, neurons, organs, flesh, and bone create my local body by constant cellular reference to the recipe of my genes, my personal DNA, which humans share almost entirely among our species, and in no small degree with all forms of terrestrial life.

The entire process of existence is in constant flux – renewal at all levels. Concerned more with doing, through interchanges among these component energies, my exterior interacts with similar material objects that are equally insubstantial. Thus, through my physical senses, which are of the same stuff, i have a conscious reaction to force, solidity, and texture. However, all i think i touch is vacuous, tenuous, energy-charged wisps of tumultuous space. What is not space is an energetic deception. Truly, it is a virtual place this universe, and you and me within it.

Our Interior

Of what is our interior? Not the organs, guts and structures that comprise the inside of the body, for that is also exterior to the true self – the psyche associating with the physical body. My interior is my formless mind. Conventionally, the term interior implies a confined, bounded space, but my interior, and yours, is limitless, boundless, and dimensionless. No scale can measure its size, though if one could, it would vastly exceed that of my exterior. Like DNA, much of my interior is the same as yours and shared by all sentient beings.

That shared part constitutes universal Sentience. A voluntary link between our nonconscious and this mutual universal Sentience resides in all of us. However, that spark of spirit is mostly masked from conscious thought. That is not to say there is any distance between Sentience and us, for there is not; it is ever-present; it permeates us. That voluntary link is a switch – the switch of realization, awareness, or enlightenment – which, when thrown, allows us profound, though attenuated access to our core being within Sentience and thence to all that there is. Most of us never throw that switch. It requires some dedication or sometime trauma; and yet, without throwing that switch, our perceived interior remains limited to our personal egocentric, superficial consciousness.

No part of my interior is private. I share all with Sentience. Within my mind resides a portion of personal consciousness and a focused node of Sentience, which, in combination, influences thoughts, represents memories, sensations, and feelings, and provides that spark of spirit. Sentience occupies no specific location, and not being bound by spacetime seems from our dimension-limited perspective to be ubiquitous and eternal. Nor does our node of Sentience inhabit a particular location in the brain, though intuitively, its main effect seems to us as focused therein, and it continually interacts with it.

The brain is more of a focal point – a place of work – for our personal conscious and unconscious mind to sensorially process and interact, particularly when relating to our body and its experiences. Personal nonlocal nonconsciousness can spread out beyond the body into the rest of Sentience. While alive, the body, through the nonconscious mind, draws down nonlocal access to Sentience and focuses it locally as unconscious input to the brain. Once the body dies, the focus dissolves as a raindrop fallen to the sea; its essence reabsorbed into the limitless ocean of Sentience.

Our Role

While Sentience and our personified versions continually exchange information, the process is usually obscured from our ordinary conscious awareness, unless we have thrown that awareness switch. Although the nonlocal portion of our nonconscious processes may appear to be associated locally with the brain; that is only because the mind is mostly a locally sponsored phenomenon. One's bodily senses provide most experiential information to the locally placed brain for processing by the mind. Thus, the local conditions informing the senses dominate the mind's day-to-day experience and apparent spatial presence.

Personal consciousness is primarily concerned with daily doing, such as surviving, seeking pleasure, avoiding pain, achieving, and ego-building. It is also concerned with thought; and at a higher level, mostly through the nonconscious connection of the psyche, it can be influenced by and reflect Sentience's creative desire to be and to become. That shows up in the oft-avoided quiet times when there can be found enjoyment in not doing, just being. These occurrences are likely to be more frequent during quiet, meditative contemplation and reflective self-enquiry. At some rare times of elective quietude, one occasionally gets to flick that awareness switch, witness oneself witnessing, and know that one is Sentience itself.

Within this context, the conclusion is that each sentient being, as an agent of Sentience, is a necessary part of what is effectively a feedback-driven evolutionary control system – an integrated physical and metaphysical complex system. The experiential feedback from these sentient agents provides Sentience with directional guidance for evolving a supportive reality and an appreciation of developmental progress. This cognitional existential system enables unbounded Sentience to transcend from an original state of undifferentiated consciousness, unconditioned by spacetime, toward one including self-aware universal consciousness. An essential aspect is that all sentient beings, conditioned as they are by spacetime, provide the necessary sensorial feedback responses to guide and permit the evolution of the universal historical existence to be sentient-beneficial, thus providing Sentience with corporeal form.

Finally, if there had to be a concise answer to the "Who am i?" question, it might be that every one of us is a local, interactive, self-aware manifestation of universal Sentience. Our purpose is to evolve and add value to the creative materialization of this emergent virtual universe.

Epilogue

Not all occurrences of awakening or insights happen during formal meditation. Many develop during workshop style contemplative sessions or gatecrash their way into consciousness during quiet moments or in the middle of the night. When asked about this apparent enigma, Shunryu Suzuki-Roshi, author of "Zen Mind, Beginner's Mind", responded, *"Enlightenment is an accident. Spiritual practice makes us accident-prone."*

The personal perspectives, insights, and speculations offered in this book may point toward a path that will eventually lead to more rational explanations of our existence than conventional reliance on deities or random chance.

Furthermore, these understandings do imply, more than other interpretations, that we and other sentient beings are more reliant on our own cognizance in creating our future existence than had been previously assumed – both mentally and physically. We, not anything or anyone else, are responsible for ourselves.

Metaphysics, Science, and Meditation

Metaphysics – meaning larger than, or beyond physics – is mentioned frequently herein and is a category of philosophy embracing the business of trying to explain the fundamental nature of existence, with emphasis on what we do not know – as in being unable to define robust manifest characteristics – and that may extend into what we cannot know.

Thus, metaphysics is mainly a label for the attempted description of the unknown or mysterious aspects of life experiences. It might be a more accessible term than, say, spirituality, which wholly concentrates on the mystic though not specifically on the theistic. However, within metaphysics are secular forms of spirituality – those that embrace the incomprehensible and intangible aspects of existence, without evoking church, deity, or superstition as explanatory belief systems. Such descriptions may legitimately remain tinged with elements of awe, wonder, and even reverence for the magnificence of our – yes, our – creation, and all it may contain, but are no longer tainted by superstitious fear and mandatory worship of mythical creatures.

As a means of homing in on a physical truth, or at least a functional representation of our existence, the scientific method is, as Einstein pointed out, the most precious tool we have. There may never be an absolute-final material truth to our physical existence, just ever-more detailed models describing ever-more tenuous evolving states. Frustratingly, all that the most successful science seems likely to achieve is to approach what may appear to be a "complete" understanding asymptotically – never quite getting there. It seems that every time we think we fully understand something, along comes an upset.

In modern-day theoretical physics, speculation may appear rampant at the frontiers of knowledge, and there is nothing wrong with that. From this melting pot of unproven ideas comes the proving process that eventually disproves most proposals. This scientific method of falsifiability, relying on mathematical proof or repeatable experimental data, is the crucial winnowing process leading toward a more robust concept or theory that can withstand the tests of critique and observation. That theory or model then remains the favoured descriptive choice until some newer one evolves with an even better record of validity. That is the essence of the scientific method.

There should be room in that process of discovery for recognizing the role of the subjective aptitudes of the mind. Few would challenge the use of individual consciousness in the process of discovery, and even among the hardiest of materialistic scientists, there is a reluctant admission that unconscious processes such as inspiration, intuition, or conscience have a valid place in the history of scientific discovery. These psychical faculties are unquestionably intangible aspects of the mind, frequently the unconscious mind, and, as suggested here, may associate with a nonlocal part of the nonconscious, and thence back to Sentience.

The secular spiritual study of existence, such as that pursued by Buddhists conducted through the discipline of meditation, may be as valid a form of subjective enquiry as is the rational pursuit of the metaphysical branch of philosophy. In combination with conventional objective science, the discipline of informed contemplative meditation could be a supplementary process for an enquiring mind to induce inspirations and important directional clues to the possible mechanics of existence or other seemingly intractable issues. These insights may in turn inform the development of objective, science-based detailed discoveries, and assist in integrating them within the holistic nature of existence. The disciplines of the scientific method and contemplative meditation may thus be employed as complementary rather than conflicting alternatives, and together may provide deeper insights than either one alone.

Conventional science is somewhat skeptical of the validity of subjective sources of comprehension due to their perceived unreliability, lack of objectivity, intermittency, and reliance on the personal integrity of the individual doing the accessing. While these are valid points, it is that very subjective access, whether deliberate or unsolicited, which has indirectly driven much scientific discovery forward. Furthermore, quantum physics is now demonstrating that all science, at least at the quantum scale, may be unable to avoid subjective influence, because of the role that entangled conscious enquiry plays in the experimental setup, observation, and reporting stages. That indirectly introduces subjectivity into the foundation of what might otherwise have seemed an objectively designed experiment. Therefore, it seems pertinent to consider more formally integrating the role of meditative enquiry into the process of the scientific method, at least as a form of inspirational tool, and relying on the more "objective" scientific methodology to furnish detail and compensate for any perceived bias in that subjectivity. Perhaps we are already heading in this direction. Many examples exist, some mentioned here, of individuals' inspirational insights and speculative theories having to wait decades before physicists can develop the experimental tools that ultimately confirm or falsify the theory by observation.

Interestingly, over the past three millennia or so, the subjective research of meditative contemplation has provided much directional, though perhaps not particularly specific pre-emptive wisdom relating to the extremes of scale. This form of subjective contemplative enquiry, though now informed by the best contemporary scientific knowledge available, may aspire to guide and leverage our creative thinking, our intuition, and our conscience in more balanced, global, and comprehensive scientific directions. Philosophy and psychology, science and spirituality, physics and metaphysics; these complementary perspectives may appear different, though when combined, they represent potential integral wisdom about "what is," just awaiting recognition. Ultimately, "what is" is both author and protagonist of our virtual universe – Sentience.

At the frontiers of physics, with some sense of unease among the more conservative scientists, there is the tentative recognition that perhaps something other than pristine physics is in play. Unidentified factors – hidden variables as Einstein called them – may influence the outcomes of what appears to be a series of highly improbable coincidences fundamental to the existence of this evolving, sentience-supporting universe. The view expressed here is that we need not look backward to theological superstitions for an answer. There is a sufficient unexplored opportunity for scientific understanding in metaphysical areas, such as consciousness and probability, to warrant developing new techniques of enquiry and comprehension. Using subjective tools such as contemplative meditation for encouraging insights that can then inform more objective analysis, with protocols to preserve the scientific method within their deployment, we may discover that consciousness plays much more than an emergent role in our existence. We may come to realize that consciousness is fundamental – all the way down!

Toward a Future

As our species travel along this journey of evolution and transcendence, our survival assumed, will we ever achieve the psycho-spiritual trail's end? Is our destiny to metaphorically hang out a shingle stating, "Perfection Achieved," then float through an enlightened, perfect existence in an unchanging paradise? Well, no – this journey has no endpoint; at least, it is indeterminate. It is a path of endless evolution, transcendence, limitless exploration, innovation, and creation. It is a path untrodden by our species. Sentience is – we are – continually pushing the evolutionary envelope. Moreover, while the concept of perfection might be a worthy goal, it has no absolute value. Like the concept of infinity, once you announce a measure of its attainment, you can always find a way to better it incrementally, and therefore it remains unattained. The concept of perfection is a relative and subjective ideal, a distant horizon never to be reached.

In several spiritual traditions, there is an inherent concept of cyclic existence, after which everything kind of resets to zero and the whole existence process starts again. Various authorities of those traditions will claim a five- or six-thousand-year cycle, some a kalpa, an inexact measure of several million years, or an eon, a billion years or more. Others will prudently avoid any actual number and refer to the ages as being an inconceivable length of time. Some of these traditions also concentrate more on much shorter personal cycles of existence through reincarnation, karma, or concepts of heaven and hell, essentially offering a much faster response time within which to receive judgment, reward, or punishment for individual corporeal doings.

Science has pegged the present age of our universe at nearly 14 billion years. On the off chance that we are halfway through its possible life, that would give an expected remaining life span of some 15 billion years of accelerated expanding existence before expiring into a cold, lonely vacuum, or some latent phase-transition version of it. More local scenarios speculate that less than 5 billion years remain before the Andromeda galaxy collides with our Milky Way galaxy, or our expanding, dying sun consumes the entire solar system, or both. Of course, we may have succumbed to an extinction event well before then, either natural or one of our own making. The latest scientific thoughts in multiverse theory also include a continuous cyclical rebirthing process. In this case, the expiring, minimal-density, final vacuum of maximal entropy undergoes a phase change of state to become the stepping-stone for the rebirth of one or more new universes. Some irony can be found in the close fit of these kind of contemporary speculations with those of thousands of years ago, albeit those voiced now being more scientifically informed.

None of these scenarios help to determine our destiny, and it is not productive to ponder on it too intensely. Still, we now have some understanding of our relationship with Sentience and the universe. Thus, we can realize that we sentient beings are already necessarily manifest thoughts – figments engaged in creating our impermanent reality. Perhaps that novel viewpoint elicits some new awareness of our likely future. One such view might be that, since we are co-creating our environment and ourselves, if warranted, we might cause a solution to the solar, galactic or universe termination issue for the benefit of all remaining sentient beings. However, before any such finality, our species may have elected to transcend this virtual physical existence… or we may just be extinct!

Armed with this newfound perspective and more profound awareness for both the physics and metaphysics of our wondrous existence, where might we go from here, and why? Just because we may realize a greater depth of awareness about our situation does not mean that we can expect radical changes in our abilities, in the various ways we physically relate to this existence, or how we live. Despite hyperbolic portrayals of historical events, we will no more be able to walk on water or perform other flamboyant miracles, than before any such awareness was realized. Our reality-coping attitudes may improve though, and with a broader perspective, we may be better able to focus our intentions, compassionately and holistically, on what matters – service to evolving existence and others on this journey. As our capacity for compassion grows, and our egocentric preoccupations with things material diminish, we will become better people. However, dramatic short-term shifts in our overt personal power and abilities are unlikely. The differences will be within.

The artist and poet, Wu Li, offered the following pragmatic observation on the personal impact of enlightenment:

> *"Before enlightenment, draw water and carry wood.*
> *After enlightenment, draw water and carry wood."*

Upon awakening, it is not necessarily the activities of life, but the quality of performing them, which changes. Given the choice of staying in a state of ignorance and stoic suffering or living the same life in an awakened state of blissful enlightenment and joy, the latter would surely be the more meaningful, fulfilling, and self-actualizing. *Nirvana* and *samsara* are "not two"; they are faces of the same coin.

The task of the enlightened one is not to rise beyond the mundane aspects of life, but to embrace them, while gently and compassionately helping self and others – all sentient beings – to evolve toward a state of enhanced awareness. It is to guide and encourage so that enlightenment is no longer a minority condition among our species. It is not to become an evangelist, not to play some spectacular, expectation-defined role, but to be a subtle beacon, unconsciously shining from within, exemplifying authenticity, compassion, and awakeness. It is to draw water and carry wood in a state of full awareness of who one is, who we are, what is and what is not, and ultimately to be able to quietly access and maintain a serene, blissful state of *nirvana* while fully functioning in this world.

Within our society are many who wish to resist change – change in things, knowledge, status, and change in relationships with others and themselves. In some cases, politically institutionalized resistance to change may appear to bring about a comfortable hiatus – a temporary stability of sorts. However, our society is undeniably evolving, as is existence. It must. Like life, it must adapt or cease; and for it to evolve, the individuals who form it must evolve. Individually, we are doing that, to some extent physiologically, though more so psychologically. Evolution requires change. Throughout the world, there do appear to be restless pockets of enquiring minds, growing in intensity. The demand for change reflects a wearying of our traditional way of dualistic, confrontational, egotistical, and superstitious thinking.

Those individuals see that, as a society, we could be so much better, and they grow impatient with the entrenched attitudes of the slow adapters and risk-averse power-mongers. These awakening individuals see value in less materialistic attitudes and philosophies, and a more integrated balance of both material and psycho-spiritual awareness. They are imbued with the perspective that both the physical and metaphysical planes of existence are their heritage and destiny. That movement is growing because the numbers of questioning individuals are increasing. This society is evolving because more self-actualizing individuals are evolving. Call it spiritual, psychological, or philosophical, but it will be individuals, not institutions, who drive this next evolution, and that is how it should be, how it must be. We may then witness an evolutionary jump akin to a metamorphosis – and a new form of awakened cultural perspective will spill out.

That next evolutionary jump of humankind may soon be upon us. It might not be physical; it will be a mental shift of perspective – psycho-spiritual, transcendental. We do not know what this will look like, and it may yet take decades to occur. Still, we will know it when, after a period of confused societal upheaval, much of humanity, individual by individual, will metaphorically split apart its materialistic mental cocoon and emerge a different thing. We will become capable of balancing multiple authentic viewpoints and cultures, realizing both our material and divine nature, and our natural leaders will reflect a genuinely cosmic perspective. It is all down to the individual, though. It is all down to us – to you, me, each one of us. Increasing numbers of individuals listening to their authentic inner voice will find their true spirit and their non-dual nature. Their linkage within this Sentience-inspired virtual universe will enable them eventually to sway the mass of humanity to depart from long-entrenched, self-inflicted suffering; to leave the pale world of flickering shadows of potential, and to emerge into the brilliant daylight existence of awakened, fulfilled, and compassionate being.

Closing Quotations

*"Believe nothing, O monks, merely because you have been told it…
or because it is traditional, or because you yourselves have imagined it.
Do not believe what your teacher tells you merely out of respect for the teacher.
But whatsoever, after due examination and analysis,
you find to be conducive to the good, the benefit, the welfare of all beings –
that doctrine believe it and cling to and take it as your guide."*

The Buddha (563-483 BCE): The Dhammapada

★★★

*"Great becomes the fruit, great the advantage of earnest contemplation,
when it is set round with upright conduct.
Great becomes the fruit, great the advantage of intellect,
when it is set round with earnest contemplation."*

The Buddha (563-483 BCE): The Dialogues

★★★

*"This Samadhi completes the transformation
and fulfils the purpose of evolution.
Now the process by which evolution unfolds through time is understood.
This is Enlightenment."*

Bhagwan Shree Patanjali (ca 2nd Century BCE): Yoga Sutras

Acknowledgements

I wish to express my loving gratitude to Hannah MacLeod (Kenzi), whose active influence set me on the path of meditation, who provided continued encouragement throughout this endeavour, and who suffered cheerfully through months of editing this book.

My gratitude also extends to the patient sisters of the Brahma Kumaris World Spiritual University, for providing structure, learnings, and challenges, during my formative stages of pursuing meditation and contemplative enquiry.

CPSIA information can be obtained
at www.ICGtesting.com
Printed in the USA
BVHW040257121021
618721BV00004B/10